Ramblings of a Chemical Engineer
Copyright © 2018 by Zaki Yamani Zakaria

Zaki Yamani Zakaria
6, Jalan Pulai Bestari 13,
Bandar Baru Kangkar Pulai
81110 Kangkar Pulai, Johor
Malaysia
Email: zaki.yz@gmail.com
Web: www.chem-eng.blogspot.com
Web: www.chem-eng.online
FB Page: www.facebook.com/ChemicalEngineeringWorld/

2nd Edition, 2019.

Perpustakaan Negara Malaysia Cataloguing-in-Publication Data

Zaki Yamani Zakaria,
        Rambllings of a Chemical Engineer / Zaki Yamani Zakaria
        ISBN: 978-967-16158-0-5
        1. Chemical Engineer.   2. Chemical Engineering
        I. Zaki Yamani Zakaria, 1976 -   II. Judul

Author: Zaki Yamani Zakaria
Language Editor: Mazura Jusoh

ACKNOWLEDGMENTS

A book lives within a community of people who support it. What a great and delightful community I have. This project commenced somewhere in 2011 and finally after 7 years, it finally materialized as a more systematic document in a form of a book.

Exceptional thanks to my wife, Assoc. Professor Dr. Mazura Jusoh for her usual, always-valuable perspective, advices and her extra role as the language editor of this book. The encouragement and insight came at just the right time.

Tarig Hussein's comments and suggestions on my first raw draft last June provided me the boost of confidence I really need to push myself towards completing this publication debut.

Special thanks to Mr. Gobal Palanisamy, a senior colleague of mine who have given me valuable inputs for improvement for the second edition of this book.

I also appreciate the constructive comments and recommendations given by Dr. Aziatul Niza Sadikin and other comrades from the Centre of Engineering Education (CEE) and School of Chemical & Energy Engineering, Universiti Teknologi Malaysia.

Dr. Siti Shawaliah Idris (Universiti Teknologi Mara), Dr. Javaid Akhtar (Universiti of the Punjab), Dr. Farizul Hafiz Kasim (Universiti Malaysia Perlis) and Dr. Thushara Subasinghe (University of Moratuwa, Sri Lanka) provided valuable comments that somehow improved this book's original flow arrangement. Thanks a lot guys.

Thanks to my talented son, Ikhwan, who assisted on some graphics in this book. My daughters, Marsya, Alya and Zara who are also there to cheer me up and motivate me in their own unique ways. Special thanks to my friends and family, and also to my parents who always support me without waver.

FOREWORD

I was inspired to choose chemical engineering when I first saw the chemical formula from my father's chemistry book. The chemical formula shapes look fascinating and interesting to me.

My father was an organic chemistry lecturer in Universiti Teknologi Malaysia (UTM). When I was 14, I read his organic chemistry book and willingly learnt from it by myself. When I was 17, I wanted to have a career associated with chemistry. Back then, my first choice was chemical engineering and my second choice was biochemistry. To be honest, I was unaware of what chemical engineers do and what the industry is like. I could not imagine it due to lack of exposure and information.

After completing my high school education, I pursued my A-Levels and took 3 core subjects which are essential for engineering: Physics, Chemistry and Mathematics. Then I continued my degree in chemical engineering. I managed to get a place in Bradford University, United Kingdom. I was unlucky because in our contract, practical training or sandwich course is not included by our sponsors. Therefore, we didn't have any valuable practical and industry exposures. That didn't matter and I kept on studying until I graduated in 1999.

Post graduate - Research and Development (1999-2002)

After completing my degree, I returned to Malaysia and was appointed as a research assistant for 5 months in UTM. I joined "Chemical Reaction Engineering Group" (CREG), where its main research was developing a single step conversion of natural gas to gasoline using zeolite catalyst. It was a very interesting topic and that encouraged me to further my chemical engineering master's degree in it. Hence, I then became a full time research student and my research title was *"Optimization of Oxidative Coupling of Methane (OCM) Using Experimental Design"*, which is part of the natural gas to gasoline research project.

Oil & Gas Exposure – Servicing Company (2003-2005)

After completing my master, I was offered a job as chemical technologist for a local oil and gas servicing company. In a year, I became a project/chemical engineer in the same company. My main task was to lead the "internal pipeline chemical cleaning" project for a local oil company. We basically have to assist the oil company to reduce corrosion activities inside the downstream pipeline and prolong the life span of it. To efficiently and effectively monitor corrosion activities in the pipelines, we utilized latest corrosion monitoring techniques such as electronic resistant probe (ER) and field signature method (FSM).

I was also in charge of the oil and gas specialty chemicals. I travelled to a number of offshore platforms in East Malaysia to conduct deoiler and descaler tests for their oil reserved. It was very challenging and fun performing those tasks. I love going offshore because the working hours are less compared to the amount of time we spent on the platform. The foods are marvellous and comparable to 5 star hotels. Entertainment and other activities such as television, movies, snooker, ping-pong, gymnasium and reflexology chair are made available for the platform dwellers. To be able to go offshore, I have to undergo Helicopter under Water Escape (HUET) training and get myself an offshore passport. With this job, I travelled extensively and visited neighbouring countries, Singapore and Indonesia, for work purpose.. In Kalimantan, Indonesia, I joined our company principal to conduct bottle test field trial for local oil company on their onshore oil rig. It was a very interesting and exciting assignment because I got to see how simple the setting of an onshore oil rig because in Malaysia we only have expensive and complicated offshore oil rigs/platforms.

Oil & Fats Industry – Refinery (2005-2008)

I love my oil and gas career but I was unfortunate because I could not continue being in that industry. The company management has bigger plans and they moved to Kuala Lumpur, the capital of Malaysia. I was instructed to transfer which I could not do because I don't want to hinder my wife's career establishment as a lecturer/researcher/consultant in UTM and we also have just purchased a house in Johor Bahru, the same year.

I seek for other jobs and managed to get one in a physical refining plant in the oil and fats industry. This is a whole new chapter and totally different from my previous job. I was required to punch in and out every time we enter or exit the factory. Life is no longer as flexible as before. I don't have ample time to do my work and that made me work longer hours and I always reach home when it's already dark. I don't really mind because it's a new working environment and I know I have to learn as fast as possible. I set my target to know everybody around my circle of work as soon as possible.

My first task given by my boss was to identify and list down all the valves in the plant I was in charge of. It was an interesting and good assignment. It made me traced the entire pipeline from the feed tank to the plant and to the product tank. I learned a lot of stuff regarding valves. I know and understand various types of valves, brand/ origin, sizes, spare parts, principle/operation, tag number etc. In addition to that, indirectly, I learned about the plant process and operation. That was just the beginning.

Being in a process plant is a perfect place to learn and put in practice my unit operation knowledge. It also gave me a better comprehension on what process control is all about. I learned about other supporting units like heat exchangers, cooling towers, high pressure boilers, utility boilers and much more. The learning curve continued every day and never stopped. Not only that I learn about all the technical stuff, handling manpower and conflict is another challenging area that I made myself good at. Manpower is not an easy matter to deal with. Some of my down line manpower never experienced any disciplinary action taken when they violated certain laws such as coming in late and simply not coming to work. Despite a series of reminder and warning, the bad attitude still continues. I could not stand it. With the support of my senior colleague, I enforced the discipline and forced them to obey. I gave the problematic staff some disciplinary action. I want them to learn some lesson and be more serious towards their responsibility and work.

During plant shut down, I learned a lot. Techniques on ensuring the fastest and effective way to cool down the plant, managing and coordinating a team of people to service the plant, conducting air test, steam test and driving the plant start-up are among some knowledge I acquired.

A Little Something from Me

There is a lot to share, but it's impossible to include everything here. I'm glad to have experienced chemical engineering in three different areas; research (academics); oil and gas; and oil and fats. Each area has their own challenges, advantages and unique. Being very vague about the chemical engineering industry during my student life urged me to improve the situation. Wouldn't it be nice if somebody can tell and share what they can expect from the industry? It will be some sort of a chemical engineering informal education for the students and other junior engineers. That is why despite of my busy life as a process engineer (previously), I progressively and continuously share some of my experiences in my *"Chemical Engineering World"* blog that I created in 2006. I sincerely hope it can provide at least some useful information for fellow young chemical engineers. I believe it's a good thing if other professional and practicing engineers out there can do the same for others to benefit. It will be a great contribution.

BRIEF ABOUT ME

I used to be a project, process and chemical engineer in the *Oil and Gas*, and *Oil and Fats* industry. I earned my Chemical Engineering Bachelor Degree from University of Bradford, UK. Both my Chemical Engineering Master's and Ph.D were obtained from Universiti Teknologi Malaysia (UTM). My research areas include catalysis and reaction engineer, renewable energy, safety and sustainability as well as engineering education.

I am a Chartered Engineer (CEng) under the Institution of Chemical Engineers, UK (IChemE) and Professional Engineer (PEng) under the Board of Engineers, Malaysia (BEM). I am also a member of Asean Associate Engineers (AAE), a Certified Energy Manager (CEM) and a Professional Technologist registered under the Malaysian Board of Technologist (MBOT). I'm now employed as a chemical engineering educator/researcher/consultant.

You can follow me through my blog and Facebook Page:

http://chem-eng.blogspot.com
(My very first blog that started since 2006)

https://facebook.com/ChemicalEngineeringWorld/
(29k followers as of October 2018)

http://chem-eng.online/
Chemical Engineering Online

MY CHEMICAL ENGINEERING CAREER TIMELINE

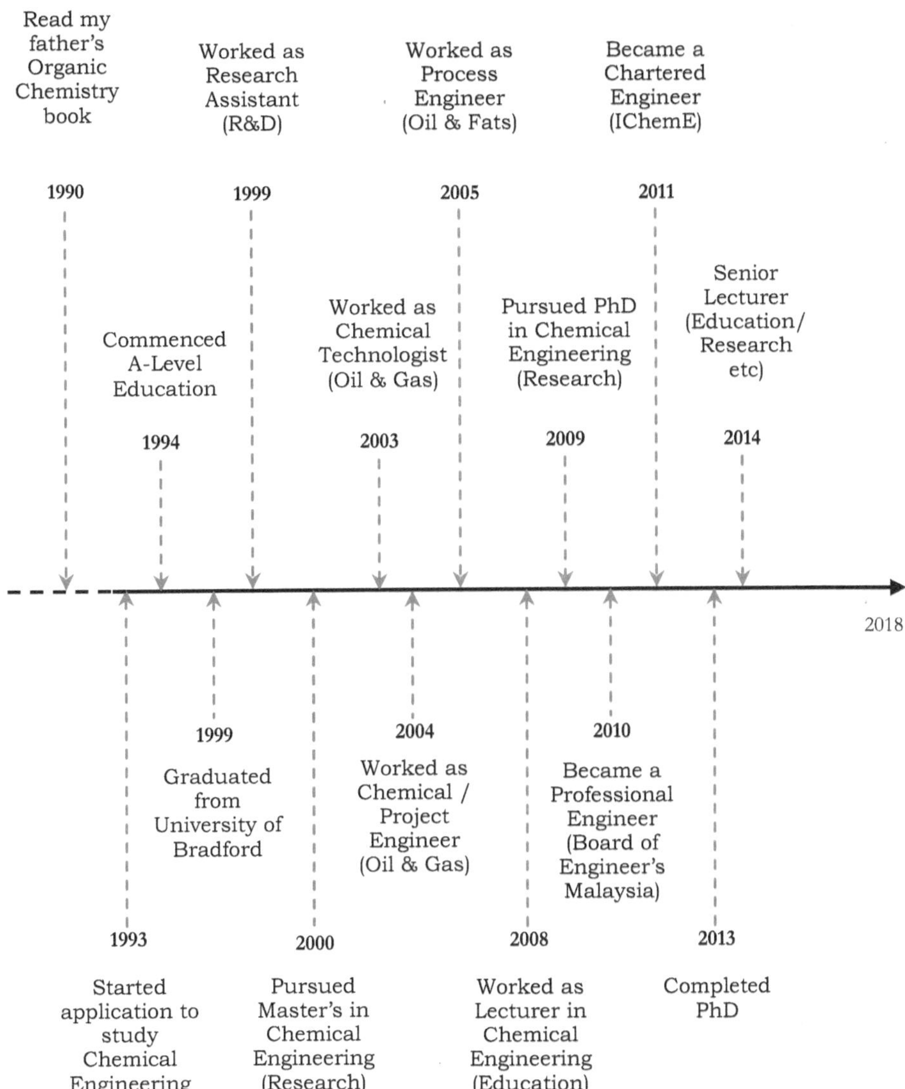

TABLE OF CONTENT

1996 - 1999

Early Chapter as a Chemical Engineering Student

When I decided to study chemical engineering a long time ago, I did not really know about the course. I fancy chemistry, and that directed me to chemical engineering. That time the internet was still not there, or perhaps very new to human being. Without really knowing what chemical engineering is, I just took the course. I applied the program from the local education ministry; got myself an interview for a scholarship and there I was on my way to become a chemical engineer, which I still did not really know about.

I started the early chapter as a chemical engineering student by entering the A-Level program at Institut Teknologi Mara, Centre of Preparatory Education. It's basically a twinning program between the Malaysian government and the United Kingdom. The plan was, to do the A-Level for 2 years, another year of first year university at the "Northern Consortium United Kingdom" (NCUK) [still in Malaysia], and then if we passed our first year examination, we shall proceed to universities in the United Kingdom (UK) for another 2 years.

A-Level is supposed to be 2 years. But in actual, the real A-Level was 1.5 years. The other 0.5 year, which was the first 6 months of the 2 years, was just pure English study, comprehension, speaking, writing etc. This is to really prepare ourselves to be proficient in English when we go to UK. After those 6 months I started learning 3 basic subjects: Physics, Chemistry and Mathematics. There was definitely some practical lesson/lab for physics and chemistry. For maths, we studied Pure Maths and Mechanics. Don't ask me now, I forgot all those stuffs. Now the objective is to score good result in the A-Level program so that we can go to good decent university in UK. To make it short, I produced average result for my A-Levels. I managed to get myself a place in Bradford University, in the Midland of Britain. But first, I have to complete and pass the NCUK before going to UK.

In NCUK program, life was not as easy as I expected. I was then a first year chemical engineering student. I have to learn more chemical engineering subjects. It was simply because NCUK is a program for first year university that uses the syllabus for a number of universities offering chemical engineering degree in the Midlands, UK. Instead of learning first year subjects of a university, we all have to learn subjects from a combination of universities, which means we have to slightly learn more. All universities offering chemical engineering will have in general quite similar basic first year subjects. So, well that's ok. I have to face it.

The subjects that I learned during that first year university:
1. Chemical & Power Thermodynamic
2. Material & Energy Balance
3. Unit Operations
4. Fluid Flow, Heat & Mass Transfer
5. Strength of Material
6. Reaction Kinetics
7. Engineering Drawings
8. Engineering Mathematics/Statistics
9. Laboratory Experiments
10. First Year Design Project

At the end of the first year in NCUK, I took the exams together with the rest of my friends. It was so tough! It was the most difficult examination that I have ever taken in my entire life. This exam was very critical as it decides whether you go to UK or you stay in the country. I can recall seeing some of my friends crying while answering the examination papers. I can also see my friends shaking their head not knowing what to answer. All papers were very tough, but the most cruel and disastrous were Unit Operation and Engineering Maths. The lecturer, to my surprise enjoyed seeing us cannot answer the questions. Oh NO! What will our future be? That question lingers in most of our heart. Are we going to UK or not?

The day came and the result was released. It's now a matter of "pass" OR "fail". Everybody was worried. Worried that they would not make it for UK. To my relief I passed the first year examination. Majority of my friends passed too. If I'm not mistaken, maybe about 10 of my friends failed and they were very miserable. I felt sorry for them. Life goes on and now I'm on my way to UK.

"I believe I can fly"
"I believe I can touch the sky"
"I think about it every night and day"
"Spread my wings and fly away"

The song by R. Kelly keeps on playing in my heart. Yes! I made it!

"Failure is the seed to success"

Kaoru Ishikawa
(Chemical Engineering from University of Tokyo,
Invented Fishbone Diagram for Cause and Effect analysis)

Mid Chapter of a Chemical Engineering Student

One of my proudest moments was when I successfully get myself a place to study chemical engineering in UK. I can remember how proud my parents were, especially my father. I was glad to make them happy. The day came and all of us, the JPA (government) sponsored student gathered at the Kuala Lumpur International Airport (KLIA) that evening. All my family members were at the airport to wish goodbye and good luck. We (my colleague and I) were very excited at this moment. We were really looking forward to enjoy the journey. Some of us have never been on a plane. So, they were more excited!

The journey took about +/-12 hours by the MAS 747-400, and we really enjoyed the journey. We had some sleep and ate some nice food served to us. We arrived at Heathrow Airport on the following morning. The temperature was very cool (autumn) and the wind was blowing strong. And yes we came prepared with all these winter clothing. We transited at London and took a Focker-28 to Leeds-Bradford airport. It was a smaller plane with a pair of engine. I don't know what type of engine. I don't think it is a Rolls Royce engine.

The wind turbulence was very fierce that time. The plane seems to be drifting up and down; left to right to left again. Some of us lost the appetite to eat the delicious croissant and orange juice served for us. But most importantly, the aircraft landed safely and we went out of the fuselage in one piece, breathing fresh midland air and it was a very cold one.

Some matured students (post graduate) have waited for our arrival and they picked us to their home. They tell us all the basic information and what we should know about the life in Bradford: how to eat, what to eat, where to buy food, where to shop, how to adopt, the bank location, the post office location, and other vital information that we really need in this new place. I stayed in a post graduate's house whom I called Abang Hilmi, with another 2 friends. It was just a temporary home before we can get a house to rent.

After we managed to get a decent small 3 storey terrace house plus a basement in the Bradford city (12, Noble Street, BD7 3BD), we moved in. It was a very cold house without any central heating and it cost us 250 pounds per month. That's very cheap for a house and a freezing one. The four of us occupied the house and we split the 250 by 4. The land lord, a Pakistani British citizen lives next to the property and never missed to ask for the rent on the 1st of every month!

Next is starting a new chapter as a foreign student.

It was a very exciting feeling to start a new semester in a new place in another country with a totally different background. The plan then was to study chemical engineering for 4 semesters (3, 4, 5 & 6), and after that get the chemical engineering degree. The subjects that i took for all those semesters were:

Semester 3 & 4:

1. Mass Transfer
2. Heat Transfer
3. Fluid Mechanics
4. Process Control
5. Foundations of Accounting
6. English Language
7. Process Design
8. Computer Aided Process Engineering
9. Process Safety
10. Powder Technology
11. Mixing Processes
12. Advance Control

Semester 5 & 6:

1. Differential Mass Transfer
2. Fluid and Particle Mechanics
3. Reaction Engineering
4. Process Control & Instrumentation
5. Applied Chemical Thermodynamics
6. Quality Engineering
7. Biochemical Engineering
8. Non-Newtonian Fluid
9. Business Strategy
10. Design Project - Alumina from Bauxite

"As engineers, we were going to be in a position to change the world — not just study it."

Henry Petroski,
(American engineer specializing in failure analysis,
a prolific author, a professor both of civil
engineering and history at Duke University)

Studying Chemical Engineering VS Practicing Chemical Engineering

In those early days, not so long ago, just few years back, we learned all these chemical engineering subjects: Thermodynamic, Reaction Engineering, Unit Operation, Strength of Material, Reaction Kinetics, Mass Balance, Heat & Mass Transfer, Process Control & Instrumentation, Advance Control, Powder Technology, Mixing, Fluid Mechanics etc. The list continues...

Well, I don't know about the others, but for me, I don't really 100% understand those subjects. The Prof/Dr/Lecturer lectures in front of the hall/room, and we tried our best (for them who really pay attention) to focus and absorb the theory of the subject. We attempted some questions or examples given to us to further establish the comprehension of the subjects. I recalled at one class, while Prof "W" was lecturing Heat Transfer, but always facing the white board, this naughty British lad creatively played with fire in the class! It was quite a scene, but the Professor kept on bubbling and facing the white board. This was a very clear indication of a very boring class. Well, boring for them, but for some Asian student and some European student, they were super laser focused.

I scored the highest mark in my "Computer Aided Process Engineering (CAPE)" under Dr Iqbal Mujtaba, a subject for semester 3, which later on the department awarded me the BP computing award that year. What I received is just a gas handbook, not a certificate. However, that's not the point. The point is I got the highest mark, but, frankly speaking, I didn't really understand the subject! And also what is Process control & Instrumentation, Advanced control?! That time, I didn't know but now after sometime working, I realized and slowly understand what I learned previously. Oooo this is this, and that is that...oooo this is actually a rotary drum filter, why can't I imagine this when I was doing my final year design project?!

We need a new creative learning approach whereby the lecturers can clearly teach and make the students comprehend what is being lectured in class.

Engineering Award

Would it be nice to be awarded handsomely for our good and brilliant effort? Would it be wonderful to be recognized by our peers and professionals from the same field? Have you been awarded for something before?

During my undergraduate studies, I was awarded "BP Computing Award 1998, University of Bradford". I received a letter during summer informing and inviting me to attend a brief ceremony that was held that afternoon (I forgot the exact date and hall. That was about 9 years plus ago…). Initially I was puzzled and wondered why I received the award. Soon, I was informed that I got the highest mark for "Computer Aided Process Engineering" subject and that led me to the award. Unfortunately, the prize was just a gas handbook. I was hoping some sort of certificate or trophy which can be displayed on my wall or desk. Well, that's fine; I just kept the official invitation letter to receive the award.

Few years after that, I was a member (research officer and master student) of Chemical Reaction Engineering Group (CREG), a research group in Universiti Teknologi Malaysia. The group (which included me) won numerous local and international awards. I even kept a local newspaper cutting about our research. The photo of the head of research group, Prof. Dr. NASA, my colleague Ir. Dr. Didi and me beside the rig was printed. It's fun and I'm proud of it. There was also documentary of our research group being shown in national television and off course I'm one of the stars in it. I acted performing some experiment to synthesize HZSM-5 zeolite catalyst (a type of silica alumina adsorbent).

1999 - 2002

What I Learned So Far

Within this short period of time, I learned a lot of things. I gained so many experiences and all are exquisite. Every day I'll learn something new and I enjoyed it. Slowly I developed my knowledge and skills in this area of chemical engineering. I will progressively and briefly share with you my chemical engineering journey and experiences.

As you grow to be a better and established engineer, you should also set specific target for your career. Everybody must have their own target. Every chemical engineer/process engineer or whatever engineer must have their own target. The target should be a big one, huge. Without a specific target in your career or life, you may just grow old as an ordinary engineer. Is that what you want?

Indulge yourself in some professional association. It doesn't matter if it is overseas or local. I am an associate member of IChemE, of UK. IChemE stands for Institution of Chemical Engineers.

To tell you the truth, although my original ambition is to become a chemical engineering lecturer, I still appreciate all my very precious experience as a project/chemical/process engineer. Not to mention also my research experience too.

*"The scientist discovers a new type of material or energy
and the engineer discovers a new use for it."*

Gordon Lindsay Glegg,
(British Engineer and Author)

My Post Graduate Degree (M. Eng)

I studied and conducted research in Reaction Engineering, specializing in Catalysts Technology. Here, I learned to synthesize catalyst for a one step process of natural gas conversion to gasoline. I enjoyed doing this research. It was fun writing the technical papers. It was interesting characterizing the catalysts after creating it from various chemicals. Our research group did won some recognitions in and outside the country. My supervisor Prof. Dr. Nor Aishah Saidina Amin (Prof NASA) is a very dedicated and hardworking lecturer/researcher/supervisor. I learned a lot from her. Together with Pak Didi (now already Profesor Ir. Dr. Didi Dwi Anggoro), we explored our research further.

"All of us, in a sense, struggle continuously all the time, because we never get what we want. The important thing which I've really learned is how do you not give up, because you never succeed in the first attempt."

Mukesh Ambani
(Bachelor in Chemical Engineering.
Chairman & Managing Director of Reliance Industries Limited.
Ranked #1 in Richest Indian List by Forbes)

My Chemical Engineering Research

While doing a Chemical Engineering Degree, besides learning all the fundamental chemical engineering subjects, we have a design project to deal with as well. Doing the design project was very interesting as we really put ourselves as a chemical engineer, thinking of how to commercialise certain project, when to get the Return on Investment (ROI), doing some AutoCAD drawing of the plant, selecting which process, working on the mass and energy balance, mechanical design equipment to purchase etc.

On top of the above, chemical engineering students will also normally be assigned to work on a small research work. This is actually a very good exposure for the undergraduate students on how to do research work. Normally the students will be allocated a few weeks or two semesters to complete their small thesis/dissertation. Some can do the research with few obstacles from the experiment etc. Some could not continue doing that research or be placed in a very advance research group which always pushes and stresses for results. Unfortunately, in my university, the chemical engineering course did not include a research subject.

For me, after my undergraduate study, I pursued with Master in Chemical Engineering. I did a full time research study in the Faculty of Chemical and Natural Resources Engineering, Universiti Teknologi Malaysia supervised by Prof. NASA. The research was actually about converting natural gas into gasoline via a single step process using modified zeolite catalyst. It was a very interesting research. I explored and learned a lot while doing the research. We synthesized and modified the zeolite catalysts ourselves. We characterized the catalyst using various instruments such as X-Ray Diffraction (XRD), Scanning Electron Microscopy (SEM), Fourier Transform Infra-Red (FTIR), Temperature Program Desorption (TPD) etc.

Satisfied with the quality of the catalyst, we tested them in our self-made experimental rig. The experiment normally took slightly more than 8 hours. At the end of the experiment, we hope to get some amount of liquid/oil/gasoline. This liquid will be analysed in a Gas chromatography to check its hydrocarbon composition. As far as I can remember, we tested the Research Octane Number (RON), and the best that we have ever obtained was 88. That test was performed with Pak Didi (my good research friend).

This outstanding research has won us numerous awards and recognitions. That was the main research in the Chemical Reaction Engineering Group (CREG) at that particular time. My research was investigating the reaction of converting methane (CH_4) which is the main component of natural gas into ethane (C_2H_6)/ethylene (C_2H_4) via the production of methyl radical (CH_3^*). This reaction will only occur with the presence of metal oxide catalyst and temperature of about 700 - 850 Celsius. I felt it was a very good research and experience. I did face so many obstacles along the way. Luckily, a new friend that suddenly appeared helped me and shared some ideas on completing the research- special thanks to Pak Tutuk.

I also could also not forget the companion of Pidun whom was doing the Palm oil to gasoline process. Also the help and support from Sean, Istadi, Kusmiyati, Rozi, Kak Ida and the rest. But, the most motivation came from my beloved wife, Mazura who supported me all the way. Thank God, it's over. However, deep inside I fancy doing a PhD in Chemical Engineering. Well, my wife was in her first year of doing her PhD, this time, specializing in Chemical Engineering > Separation > Crystallization engineering.

"This job is a great scientific adventure. But it's
also a great human adventure."

Fabiola Gianotti,
Higgs Boson physicist

Design of Experiment for Chemical Engineering Research

For those chemical engineering post graduate students whom are doing research, or who might plan of doing research, planning your research is very important. There will be a lot of experiments to be carried out. There must be a systematic way of doing all the tonnes of experiments. And one technique to do so is by using the **Design of Experiment (DOE)**. My wife and Pidun was going to use this tool for their pH.D research. I did use this as well during my Master by research work few years ago.

DOE is a very useful technique to plan our research. From DOE we can optimize the research and know all the processing parameters that can yield optimum result. So what is actually DOE?

In an experiment, we deliberately change one or more process variables (or factors) in order to observe the effect the changes have on one or more response variables. The (statistical) design of experiments (DOE) is an efficient procedure for planning experiments so that the data obtained can be analysed to yield valid and objective conclusions.

DOE begins with determining the objectives of an experiment and selecting the process factors for the study. An Experimental Design is the laying out of a **detailed** experimental plan in advance of doing the experiment. Well-chosen experimental designs maximize the amount of "information" that can be obtained for a given amount of experimental effort. The statistical theory underlying DOE generally begins with the concept of process models. Hence, it is great if you can master this DOE technique in order to optimize your experimental research.

Process Models for DOE

It is common to begin with a process model of the `black box' type, with several discrete or continuous input factors that can be controlled, that is, varied at will by the experimenter and one or more measured output responses. The output responses are assumed continuous. Experimental data are used to derive an empirical (approximation) model linking the outputs and inputs. These empirical models generally contain first and second-order terms. Often the experiment has to account for a number of uncontrolled factors that may be discrete, such as different machines or operators, and/or continuous such as **ambient temperature** or humidity.

There are various software/packages that offer DOE. Among them are **Statistica, Design Expert and DOE Fusion.** So, in case you're interested to learn about them, get hold of the software. You can start learning from the help or **tutorial** available in the software. But, learning by yourself will take a huge amount of time. I know because I experienced that before. It's better for you to go through and then ask somebody who knows DOE. That will be a better way of learning. But if you/or your research group have the budget, request from your boss/supervisor to go for a dedicated course to learn DOE.

OK, to go a little bit deeper, there are a few designs that can be chosen under DOE. There are **Plackett Burman Design, Response Surface Methodology (RSM)** and others. I used RSM before and it was very useful. I can have a lot of analysis generated in front of me by using RSM. It's a really fantastic tool.

2003 - 2005

Chemical Engineering Career

In 2005, I started writing a blog to document and share all my experiences. The following is my first post in my *Chemical Engineering World* Blog.

Welcome to my weblog. I just want to share some experience/knowledge that I collected so far as a chemical engineer/project engineer and currently as a process engineer. Before that let me introduce myself.

My short name is Zaki. I obtained my Chemical Engineering degree from Bradford University, UK in 1999. Later I did some research in Reaction Engineering in the Faculty of Chemical and Natural Resources Engineering, University Teknologi Malaysia (UTM). After a while, I pursued my Master's Degree, studying Chemical Engineering (Full Research) at UTM. My research title was Optimization of Oxidative Coupling of Methane using Design of Experiment. It's a very challenging subject to do. I survived it.

After completing my master's degree, I joined a local oil and gas service company, doing various types of oil and gas activities. With this company, I travelled a lot, here and there, on-shore and offshore, local and over-seas. I enjoyed this job and the pay is good too! However, the good thing did not last long. I have to accept the fact that the management's decision to move the company to Kuala Lumpur is unavoidable. Having my family and my life in Johor Bahru, it was impossible for me to leave Johor Bahru. I had no choice. I have to search for other job.

I went to this interview in a refinery in Pasir Gudang. I was accepted as a process engineer and my task was to take care of one of the largest refinery plants in Pasir Gudang and some projects and utility jobs. Here I learned a lot.

Preparation in progress for pigging for the 48" Gas Condensate Line

Having a chemical engineering degree, you can work in various interesting and challenging field. You can be working as a chemical engineer, process engineer, project engineer, chemical specialist, researcher, lecturer, consultant, oil & gas field etc. For me, I've been a research officer, worked in the challenging oil & gas field & presently working in an intense oil processing refinery. And I am really proud of my tasks and responsibilities.

"If you don't have a competitive advantage, don't compete"

Jack Welch
(American retired business executive, author, and chemical engineer.
He was chairman and CEO of General Electric between 1981 and 2001)

My Intention

When I first start writing (in this blog), my intention was to provide more information about the career as a chemical engineer and any career closely related to it. This is mainly because; I was also not sure what chemical engineering is. I did mention in my previous post something about how I have vague vision about the field of chemical engineering, but I decided to pursue a degree in this discipline because I like chemistry. Boy I was wrong.

Therefore, I would like to give all information related to chemical engineering to future chemical engineering students. I would also like to expose my experiences as a practicing chemical engineer to those chemical engineering students, so that they will know more or less how and what does a chemical engineer do. What can they expect as a chemical engineer? Maybe I'm not that old yet, and do not have that huge experience in chemical engineering stuff, but I believe I have been exposed to various fields of chemical engineering after I completed my degree.

After completing my chemical engineering degree from University of Bradford, I returned to my country, Malaysia. My ambition was to become a lecturer. I began to work as a research assistant and pursue my master's degree in chemical engineering specializing in Reaction Engineering - Catalysis Technology. I was there doing research work for about two years.

Then, after my master's degree I joined a local oil and gas service company. In this company, I handled specialty chemicals associated with oil and gas. It was really exciting, interesting and challenging. I've been with that company for about 2.5 years.

I moved on with my life and got a job in a palm oil refinery in Johor Bahru. However, I have to commute daily for about 98km in order to go and return from work. This is another chapter of my chemical engineering career. This was a challenging and interesting one. At the point I'm posting this story, I've been working in the refinery for about 1 year and 3 months. However, the amount of knowledge and information that is kept in my CPU brain is so much. Luckily the RAM is also high☺.

From my experiences that I gained all these years, I hope I can help few people who chose chemical engineering as a career some information and knowledge. I hope this can further enhance your comprehension and perspective on chemical engineering.

I'm also working on to create another page dedicated to the subject of heat exchanger, cooling tower, boiler, specialty chemical, instrumentation, automation, oil and gas. However, I guess that will take some time to complete as I'm also working full time. I can only do this after work when I have the inner energy and time.

"Engineers ... are not superhuman. They make mistakes in their assumptions, in their calculations, in their conclusions. That they make mistakes is forgivable; that they catch them is imperative. Thus it is the essence of modern engineering not only to be able to check one's own work but also to have one's work checked and to be able to check the work of others."

Henry Petroski
(American engineer specializing in failure analysis,
a prolific author, a professor both of civil
engineering and history at Duke University)

Chemical Blending

When I was working in the oil and gas servicing company, doing some oil and gas business, we took care of several pipelines which transports petroleum condensate and gas. The pipelines were quite long, several kilometres. We have to maintain the pipelines from corrosion threat. Therefore, we have to select suitable specialty chemicals such as degreaser and corrosion inhibitor, to be injected in the pipeline. The degreaser main purpose is to provide enough alkaline so that any debris or sludge deposited in the pipeline can be easily remove by using the bi-directional pig (bi-di pig).

The corrosion inhibitor which is an alkyl ammine base (Quaternary pyridine alkyl ammonium), a water base concentrate will create a thin layer on the internal side of the pipeline to protect from any corrosion to develop. After injecting the corrosion inhibitor, a filming pig will evenly distribute the chemical onto the internal pipeline surface.

The petroleum condensate pipeline is a 10" diameter pipeline whereas the gas pipeline has a diameter of 48". So, can you imagine how big is the bi-di pig and the filming pig for the gas pipeline?! The weight of each 48" pig is 2 tonnes. It's not an easy job to do the pigging job for the gas pipeline. However, the challenging task for me was to prepare and blend the chemicals to be injected inside the pipeline. It's a very tough job too. I have to blend degreaser up to 15 m^3 and corrosion inhibitor up to 3 m^3. It's a lot amount of chemicals to be blended using just some 2.7 m^3 poly tanks at the site, in-situ. I have to coordinate all the blending process, manage the transportation of water, and be aware of the timing and also supervising the process of injecting the chemicals into the pipeline.

But, the most important task was to ensure the chemical's quality. I have to conduct several tests before allowing the chemical to be dosed into the pipeline. I have to conduct compatibility test, foaming test, emulsion test and few other tests. If the test failed, I have to adjust the chemical blending accordingly until the quality is obtained. All of these chemical blending processes took place on site under the sun! Imagine working under the sun wearing the nomex and tyvek coverall (You will read about this safety coverall in page 29). Uhhh! It was so hot.... Just imagine that I can easily drink and finish the 500ml mineral water bottle in 8 seconds!!! Yeahh...it's that hot....

Pipeline Maintenance

Take care of your pipelines properly. Give them a good care. Never ignore them if something out of ordinary occurs. Previously, I have a job of just taking a good care of pipelines. Why we have to take care of these pipelines? Why some oil and gas companies will invest a lot on taking care of their pipelines? Well, it's simple. These pipelines transport very precious oil and gas. Hence, we must protect these treasurable products.

Then, the next reason is that it is not cheap to construct and lay pipelines, especially for a very huge distance. Therefore, we want to really protect the asset (the pipeline). We don't want to keep on constructing the pipeline after a just 5-6 years of constructing them.

I helped handling the pigging activity in 2 pipelines (10" and 48") covering a distance of about 14 km. The main pigging activity is to create/put film of corrosion inhibitor into the internal part of the pipeline. This is to protect the pipeline from corrosion developing. Before injecting the corrosion inhibitor with the filming pig, we have to inject the pipeline with some amount and some concentration of degreaser, to remove any debris to scrap of some unwanted physical material trap inside the pipeline. Well, selecting the pigs and also the chemical is critical. We cannot simply apply any chemical as if it failed, we can anticipate a problem in the downstream section.

"Try not to become a man of success,
but rather try to become a man of value."

Albert Einstein
(Influential theoretical physicist)

Helicopter Crash - Tapis B Offshore Platform

The helicopter crash at offshore Trengganu (while attempting to land at Tapis B Platform in a storm) last Sunday (5/11/2006) really caught my attention. It was a Super Puma helicopter which transported 19 passengers + 2 pilots (The pilot did not survive). I still remember being on a Super Puma when I travelled offshore long time ago. I did not favour being inside a Super Puma. My friend, Tom, also prefers not to travel in it. We don't feel comfortable being in it. The control, maneuverer and stability are the main reason. I rather choose to be in a Sikhorsky 60 or Sikhorsky 76. Well, that was a long time ago. But the importance of the "Helicopter Under Water Escape Training" is really vital for this situation. It also reminded me that being in the oil and gas - offshore industry is fun, interesting, challenging and sometimes dangerous.

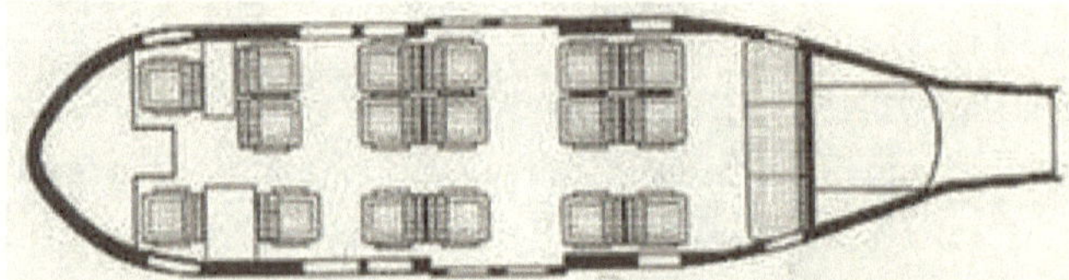

Sitting arrangement inside the helicopter

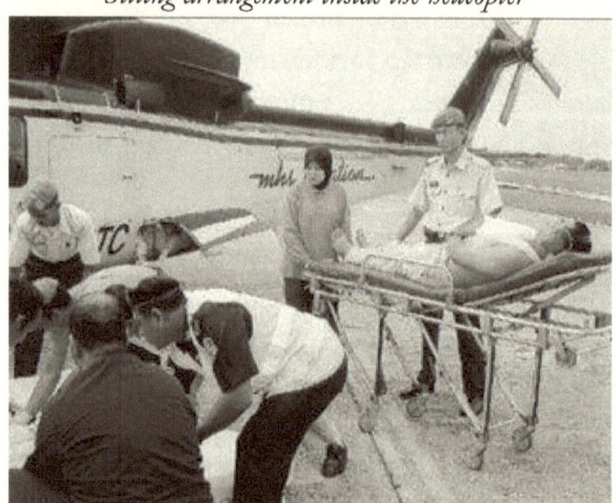

Rescue workers from the Civil Defence Department transporting the victims, oil rig workers and the co-pilot, to the Kuala Terengganu Hospital yesterday. — NST picture by Imran Makhzan

Going Offshore

I bet some of us have the desire to taste and feel the offshore life. For me, it was simply an unforgettable memorable wonderful experience.

If you want to be working in an offshore platform, you should consider working with an "oil company" (Examples: Petronas, Shell, Exxon-Mobile, BP etc.) OR "oil and gas servicing company" (Examples: Weatherford, Schlumberger, Halliburton etc.). Actually, there are many oil and gas servicing companies out there. The ones that I mentioned above are just a few of the international service companies. I previously work for a small local oil and gas servicing company, specializing in specialty chemical optimization for the oil and gas industry. I must say, I am proud and I enjoyed working in the offshore platforms.

Before you can go to any offshore platforms, you must participate and pass a training called "Helicopter under Water Escape Training" (HUET). The training is normally conducted for three days. One of the training modules requires you to escape from a simulated helicopter crashing in water (swimming pool) in various ways. Another module trained you on fire fighting and fire escape. All the trainings are really physical and challenging. Therefore, prior to attending the training which costs about MYR1400.00, you need to undergo a complete medical check-up specially designed for offshore and marine people.

These are all compulsory to ensure those who work at offshore platforms to be knowledgeable and aware of how to help themselves during emergency situation. Ability to swim is a huge advantage.

Going Offshore II

Previously, I've explained about some preparation on going offshore. There are actually a lot more to do. After getting the Helicopter under Water Escape Training" (HUET) certificate and passing the offshore medical checkup, you have to make an offshore passport. To obtain an offshore passport, you need to register with an oil company (Examples: Petronas, Shell, Exxon-Mobile, BP etc.). However, it's not that simple to get them. You won't get an offshore passport for fun. You must either be working with the oil company or an oil and gas servicing company which has a job or project in an offshore platform, in order to get the offshore passport. Bring along 2-3 photos depending on which oil company you're registering.

Without the offshore passport, you are not allowed to work at an offshore platform. Some oil companies allow people to visit an offshore platform without an offshore passport for a one day visit only, depending on the importance.

Once you have the passport and a job at an oil platform (you must already know the date when you're supposed to travel to that particular platform), you need to book a seat on a helicopter from the platform's Radio Officer (RO). The RO will then book the flight for you and inform the schedule of the helicopter. This was the procedure at my time. I am not sure on the present procedure.

"Make mistakes, but never repeat them"

Harsha Bhogle
(Chemical Engineering from Osmania University, Hyderabad.
Successful and renowned Sports Commentator)

Going Offshore III

After knowing your destination (offshore platform), you must go to the helicopter airport and check-in. At the counter, you must present your identity card and your offshore passport to the officer, besides informing your destination platform. The officer will then inspect the documents and check your flight booking made by the RO from the platform you're going to (you must liaise with that particular platform to do the flight arrangement prior to the departure) in the computer.

If your flight booking made by the RO existed in the system, you'll receive the boarding pass. Your luggage will be weighed and it must not exceed the allowable limit. The luggage will then be collected and sent directly to the helicopter after an x-ray and thorough inspection have been made. No flammable or explosive materials allowed.

"Luggage Tag" must be attached neatly on your luggage. The tag is important because it shows which platform the bag will go to. You must put on the correct tag in order to make sure that the luggage will arrive at the platform together with you. Some examples: Platform Bekok C will have a tag: "BEC", Platform Dulang B will have a tag: "DLB", Platform Tiong A will have a tag: "TIA". For your information, there are many platforms out there (South China Sea) and you must not make a mistake attaching the tag. The luggage will travel in the same helicopter as the passenger.

Mobile phones are also not allowed. You must leave your hand phone behind either by renting a safe box (this service is provided by the airport) or leave it inside your car or ask somebody to keep it for you. I always leave my hand phone inside my car parked at the airport. Again, this was the procedure at my time. I am not sure on the present procedure.

Going Offshore IV

After receiving the boarding pass, you have to go into a room to watch a safety video. The video will show some safety measures to be taken while being in the helicopter and the offshore platform. After viewing the video, you have to wait to be called to board the helicopter.

Most of the time Sikorsky 76 helicopter will fly you to the offshore platform. If I remembered correctly, I flew with Super Puma one time only and it was a really frightening experience. However, according to my colleague, I didn't look bothered and I was sleeping most of the time, including when the chopper was having some rough time. If only he knew how I felt at that time.

As the helicopter took off, you'll be heading to the sea. Initially it will feel wonderful (especially for first timers) because you get to fly in a helicopter and everything (houses, vehicles, etc.) looks tiny from above. Then, you'll fly above the beautiful blue ocean. It's nice, pleasant and very peaceful. Unfortunately, the beautiful blue ocean scenery remains the same throughout the whole journey and it can be boring. That is why I prefer to sleep after I got enough of the ocean. Normally, after approximately one hour (sometimes 1.5 hours or 2 hours), you'll see a platform and the helicopter will decelerate and descend. You will see the helicopter pad and the platform's name largely and clearly visible from the helicopter. "Bekok C" was the first offshore platform I visited and it is a gas gathering and processing platform. It was certainly a moment I won't forget.

Safety Definitely MUST Come First

My previous job which is in the oil and gas field was very strict in safety and health. We cannot tolerate safety. Safety is the top most priority. We must wear all the required personal protective equipment (PPE). If the PPE is expensive, your boss/company must be willing to spend on them (Many of those PPE are very expensive) If not, you cannot perform your job. The job requires all sorts of details. Before performing any job, we must have a pre-job meeting. In the pre-job meeting, documents such as Job Safety Analysis (JSA), Method of Statement (MOS), and Emergency Response Procedure (ERP) must be clearly defined and presented. Then only if everything is fine and accepted, we shall receive the Permit to Work (PTW).

So, there we go, do our work, wear safety helmet, safety goggles, Tyvek coverall (protecting from chemicals/ oils/condensate), safety boot, rubber glove and cotton glove. After working for some time wearing those PPE, we found out that the oil/condensate that we dealt with contained traces of mercury. So, guess what? The regulation became stricter. We have to discuss things in more detail in the pre-job meeting to overcome the hazard from mercury. Some additional steps/precautions have to be taken. In addition to that, it was also discovered that the flash point of the condensate that we dealt with during work may ignite fire (or worse, explode). Hence, again additional discussion...bla...bla... And more precaution...more PPE.....

Finally it was concluded that we have to wear two (2) coveralls at one time during work, one is NOMEX - a fire proof coverall which can protect us for getting enough time to escape if we are caught in fire and the second one is Tyvek coverall that is mercury/heavy metal proof. Gosh! Could you imagine wearing 2 coveralls under the hot sun while doing our pipeline maintenance work?

Not only that, we have to wear 3M full face piece with mercury filter - 6009 and also organic gas filter 6003....OOooo man...it's damn hot inside the coveralls. I'm sweating like ##$%%*!#? It's very hot.

All of these scenarios clearly show that safety cannot be compromised. Safety must come first in whatever means.

2005 - 2008

Not My Typical Day as a Process Engineer

At this point of time, I was working as Process Engineer at an international oil and fat company. A lot of things happened this particular week, within these two days. In the early hours today, electricity power was out. The whole refinery was affected and the whole plants have to stop. This was already a bad sign of the day, I guessed. I whispered to myself.

Then we realised that it happened for a matter of seconds only. The power provider had switched the power station and unfortunately that resulted to loss of electricity power for 2 - 3 seconds. The effect to my plant was a massive downtime.

During this sudden power breakdown, many pending maintenance job that could not be performed during normal plant operation was carried out, immediately. But, the timing must be monitored. If the maintenance work took a lot of time, and the time to regain power is less, then the maintenance work should not be carried out. However, if the work requires less time, then it might be possible to decide on doing the job fast. This is the point where decision making is vital. And it needs to come as soon as possible.

My colleague installed the uninterrupted power supply (UPS) system (a power back up system) for the PLC (programmable logic controller) in my plant. The process took about 45 minutes. They cannot install the UPS during normal plant running hour because to install the UPS, the PLC must be switched off (the plant is running 24 hours). So, when the plant is down, it was a fantastic opportunity to install UPS.

As soon as the power returned, the plant supervisor and operator were very busy doing their job starting up the plant. Many sections need to be handled and be taken care off. The vacuum system needed to be stabilized. The high pressure boiler needed to be switched on to obtain the desired pressure to reach targeted temperature. The cooling tower needed to be in great shape to supply water to plant to establish good vacuum.

Unfortunately, one of the cooling tower fan assembly was broken when we attempted to start up the plant. So, we cannot use that cooling tower. We have to use another cooling tower which was under cleaning. Luckily, cell no 1 of the cooling tower which was under cleaning was ready for duty. However, the 12" gate valve on top of that cooling tower was leaking and had to be replaced with a new one in order to make the cooling tower work. Another 1.5 hours there. Soon, a crane entered the plant and lifted the damaged fan down for maintenance. The 8 bladed-fan was serviced, repaired and fixed backed to its original position.

At the same time, during this plant start-up, many unexpected events can also happen. This is one of the most critical parts of running a plant. Any careless mistake can lead to a devastating disaster. One of the vessels was caught on fire. Luckily, it was detected fast and everything was under control.

Then, when we thought that everything was going to be ok, suddenly the power was cut-off again (for the second time within just 3 hours from the first one). I immediately called the *chargeman* and asked him to check out and rectify the problem. It was not only a problem for my plant, but also the entire plants in the factory again, which now included the utility boilers. This cannot be good. This is really bad. This happened near the lunch hour time. I got myself some update of what really went wrong by visiting the LV room. It was found that the main fuse switch for all the plant was totally wiped out! It looks really awful, toasted, and black in colour! It burnt and nearly became like coal. Estimated time to fix the mess was 2 hours. Hence, we had no choice than to just wait for another 2 hours and we did some other maintenance works that can be done while the power was gone.

After 2 hours, the power returned. Now, another problem appeared. The boiler cannot be started because its transformer suddenly cannot work. That piece of "plastic and metal" need to be replaced. After that, the boiler can be started and steam was delivered to respective plants with sufficient pressure.

At the same moment, I must make myself available for an appointment which was arranged a week earlier with a supplier who came from Kuala Lumpur. We were discussing some stuff about networking, IT and some PLC connection matter. This was already another big subject. The problem now, I was called for a sudden big meeting with the HQ. Oh no! What now? I had to inform my senior that I cannot join the meeting. I have to discuss something with this supplier. We already had an appointment.

Earlier before lunch, I was called to attend the daily production meeting. Usually my senior will attend the meeting, but on this day, some visitors from another country came and he had to lead the tour guide in the plant. So, there I went for the meeting and of course I was asked a few things about whether my plant can run the type of oil that some customer requested?! When? Tomorrow! OK. This is not normal. We have not run this oil before, so I cannot answer it straight away. I must refer to my senior executive and senior supervisor to check on the feasibility of the process.

Another project that I handled was the auto tank gauging (ATG) which had been experiencing some communication problems. However, the power problem unexpectedly made the communication improved and the system became OK. As I prepared myself to go home, I noticed that the earlier ATG problem came back. Oh NO! I shook my head in disbelief. It's happening again.

All of these happened in one very long day today and as always, as I reached home, it's already dark. I cannot even see my flowers blooming during daylight because I always reach home when it's already dark. Luckily tomorrow is a public holiday☺.

High Pressure Boiler in Plant

For any plant that requires very high temperature (>200ºC) in their process system, they need a high pressure boiler. This high pressure boiler can provide you with the high temperature that your plant requires. Operating and maintaining the high pressure boiler requires specific knowledge and some experience as well. For me, I still need to learn more about the high pressure boiler. I also need to have more practical exposure on how to start, trouble shoot and operate the high pressure boiler. For now, I do know a little about the high pressure boiler, its operation and maintenance. A friend of mine, yesterday, asked me about his problem on operating a high pressure boiler. The following is his email to me.

"Hey, I want to check with you on something. Have you experienced high pressure boiler trip due to low water level in your plant? As I know, it is a closed system where all water vaporizes to steam and condense into condensate in the system. At my place here, the HP boiler tripped due to water low level. But, they cannot find the leakages"

Well, what my friend was facing sometimes happen in any plant using a high pressure boiler. When there is some water loss from the high pressure boiler and it is a closed system, there's definitely a leakage somewhere. You need to search, find the leaking point. Get some more help. Get more people searching for the leaking point. Do a steam test to easily detect the leakage. When you find the leakage, treat and seal the leakage point. This depends on how bad/how big/small the leaking point is. Just rectify the problem, and make sure there's no more leakage inside your high pressure closed system. When there's no leakage, water loss would not happen. The water will just follow its cycle from being a steam, then condensate, and steam again, and condensate again, and so on...without disappearing....

Note...I'm not really a high pressure boiler expert. I also need to learn some more.

Auto Tank Gauging Project

One of my tasks as a process engineer is to lead/handle the "Auto Tank Gauging" (ATG) project. It's a project that is very challenging. I faced a lot of obstacles that I have expected and also never expected. What is this ATG project? Let me brief about it.

There are more than 70 oil storage tanks at the plant/factory that I worked in at this point of time. Every morning, the operation department/pump house boys will do the manual dipping / ullaging to measure the oil level inside the tank. From this, operation department can know the stock of the oil. The process was so far ok and the oil level readings were fine.

However, in order to constantly monitor and know the oil level and tonnage of oil inside the storage tank, ATG is introduced. ATG stands for Automatic Tank Gauging. It will basically measure the level of oil inside the tank using micro pilot radar / ultrasonic wavelength and/or using level probe. By using this instrument, simply said, we can see the oil levels inside the pc that is connected to the instrument at the tanks. Besides the instrument to detect the level, temperature transmitter is also installed. This is vital to measure the temperature of oil at all time. The temperature will affect the density of oil. By knowing the oil level (height), we can get the volume of oil inside the tank. Having the volume of oil multiplied by the oil density, we can get the tonnage. The tonnage is the most crucial information for the daily oil stock.

The whole process of installing the instrument, laying the cables, dealing with suppliers/contractors, commissioning, troubleshooting problems are tough. It really requires a very good coordination. Not much of my chemical engineering knowledge was put on practice while handling this project. It's more towards managing and also learning some SCADA system program, monitoring the tanks and also fine tuning the instrument to get better accuracy.

I experienced a lot of technical problem while handling this project. The lists of problems are so long that I think I cannot write it here now. It may require a lengthy write up. Imagine more than 70 storage tanks. Imagine each of the storage tank have their own issues, challenges and problems. I think it's good to share the experience to others when the time comes. That is when the project is totally completed.

3D Trasar - The Cooling Tower Might be Cool with it

As I mentioned in my post earlier, handling a cooling tower might not be that cool if the cooling tower is creating so much problem from so many angles. The problem in cooling tower triggered from other sources will create massive operational and process problem. Operating costs will also increase. Maintenance costs would also definitely be there in the calculation. You would not want your cooling tower to have a very deviating pH. You also don't want your cooling tower to have so many Legionella bacteria cultivating inside your cooling tower. And you don't want any chemistry to be not right in the cooling tower water basin. You want your cooling tower to be in a tip top condition, providing desired temperature for the process plant.

Most of the current and previous method is based on the "dose and hope for the best system" which means the chemical supplier comes and do some technical checking/evaluation of your cooling tower and its water quality. Testing some water parameters, take down some notes and provide you with some black and white report. The question is how frequent will he do this? Once a week? Once a fortnight? Who knows what the water quality is when nobody is checking it in between those supplier visits?! What about the lab? Is the lab checking the quality on a daily basis? If yes, what about the interval between those lab sample testings? Perhaps your cooling tower is slowly deteriorating and suffering from some "alien" attack!!!

How to overcome this? In early 2006, I knew that Nalco have come out with their secret weapon in having a better control of the cooling tower. Nalco took few years to develop the best control system to overcome stresses inside the cooling tower. They developed 3D Trasar, a tool to monitor and have better control of the cooling tower. It's simply a very good and impressive invention. Presently, many company already installed the 3D Trasar inside their cooling tower. One good thing about 3D Trasar, you can monitor and control the cooling tower from anywhere in the globe. Enough about that, I'm not the expert of this 3D Trasar. You better get further information from Nalco. Bottom line is that, having such continuous monitoring and controlling tools for our cooling tower is going to be a great investment, especially in terms of cutting down cooling tower maintenance cost and the water treatment chemicals.

What You Don't Learn About Plate Heat Exchanger in Uni

For those chemical engineers who are new or chemical engineering students, or chemical engineers who have not yet experienced dealing with or operating a plate heat exchanger, well, this plate heat exchanger (PHE) is not as simple as what we see in those any heat transfer text book or Perry's.

Nobody teaches you how to take care, operate, change plates, change gasket of the PHE. You don't get this information in the text book. The text book only has some theories and examples of calculating the overall heat transfer coefficient, log mean temperature difference (LMTD) and few other formulas.

You'll ONLY get the knowledge went you deal with it, or learn it from your senior engineer, or from your supervisor or fitters/technician. We were also not thought of how long is the life span of the PHE right? How long is the time where the PHE can work with highest efficiency of doing its main job of transferring heat between two liquids? How many plates are we supposed to allocate for a heat exchanger? We never learn all those stuff in university. We learn them went we deal with them. It's quite difficult to discuss about the operation and other related issues regarding the PHE. It will be a long discussion.

The most basic thing about PHE is that we want to "exchange heat". We want cold fluid to reach certain desired temperature higher than before and hot fluid to reach a lower designated temperature. Well, sometimes you don't get the desired temperature too. There are many reasons for that. Check the flow rate. The flow rate affects the final temperature. Check the fouling at the plate and frame. Is it serious? Is the heat exchange efficient? Wow! There are many things to consider and look into in this one single important unit operation equipment.

Plant Shut Down - The time to explore

Sometimes, we have to face the fact that a sudden plant shutdown is unavoidable. It maybe because of some stupid but valid reason. Some repair works have to be done. Or some additional piping must be carried out to improve the process system. The list continues. During a plant shutdown, all jobs that can ONLY be performed when the plant stops, will be rushed.

During my first plant shutdown, I learned lots of new things. My learning curve suddenly increased. My understanding on this plant became better and better. I know some other people have their own experiences dealing with this shut down. Well, during this time, all of us, technician, fitters, maintenance people, production people, operators, contract worker/helpers will combine and work together to repair/clean/rectify any problems. Some leaked valves will be replaced. Some identified cracking pipes will either be replaced or welded. Maybe some heat exchanger needs some RTD or temperature sensor to be installed upstream and downstream. One or two heat exchangers that don't have its substitute might also be serviced. The list of shut down jobs continues...too long to list it here....

During this period, I took advantage of getting inside some of the vessels. I travelled deep inside all vessels that were possible for me to enter. It was hot and dark, and we need some fan ventilation to help us breath. We checked the condition inside the pack column, deodoriser etc. It's dirty, oily and slippery. Imagine the temperature inside this vessel when it is in operation 260°C or more. That's pretty hot inside.

For me, the plant shutdown projects are some activities which are out of our daily routine job that improve our comprehension of the plant processes. What do you think?

I was checking out the manhole of the deodoriser. Soon after that I was inside the deodoriser. It was dark, hot, slippery, very small for human to enter, definitely not a place to be but we need to check, clean and service it.

"Science can amuse and fascinate us all, but it is engineering that changes the world."

Isaac Asimov,
(American writer, professor of biochemistry)

Cooling Tower is sometimes not that "cool"

Handling cooling tower when it's ok is such a great feeling. Knowing how to balance between the operational, mechanical and water chemistry is what we want, and when the cooling tower water produces the sort of desired temperature and water quality is ok, everything should be in control.

But what happen when suddenly everything is out of control. When the water quality is worsening, turning brownish or greyish or whatever colour you can mention. When all heat transfer in your heat exchanger is no longer operating at its best capacity due to ineffectiveness of the heat transfer - "fouling" or other reasons. When your chiller gets choked! Suddenly the cooling tower is deteriorating. All sorts of greyish slimy things appeared sticking on the infill. The look of your cooling tower at this point is tremendously horrifying. At this point, the water in your cooling tower basin is producing unwanted blackish foam. The sand filter is suddenly choked in a very short time. Back washed?! How long? No matter how long you back wash, the blackish, greyish colour water keep pouring out from the 3" mild steel pipe of your sand filter. What the hell is this?! This is not what I want! Oh man....my boss is going to screw me up.

The pump is delivering lower flow rate than usual when you check the flow rate. Oh no....my vacuum system is on its neck! Oh no my final oil product is too hot to be stored in the storage tank. Everything is not right at the moment.

Now, whose fault is this? The supervisor? The plant operators? The utility engineer? Production exec? Or is it me? Who? You keep shouting to yourself deep inside. Oh no! Is it the water treatment chemical contributing to this disaster? Do some WHY-WHY analysis to outsource its origin. Your headache is skyrocketing exponentially suddenly. Some oil might be leaking in any of the heat exchanger or chiller in operation. But which one? From which plant? Or no! This is totally a complete combination of disasters.

Suddenly your cooling tower is under stress! Settle the problem ASAP. How? Go and check all your lines from cooling tower. Check all your heat exchangers. Check everything. Something is totally penetrating the cooling water line (which is a closed system). Do the air test, hydro test, take some sample from all the water inlet and outlet of your heat exchangers, chillers and check which one has the highest concentration of oil.

Damn this hydrocarbon (oil)! Constantly feeding the legionella filming bacteria which is sticking on the internal wall and the entire cooling tower infill. And the search for the leaking point continues. It's not easy. Keep on searching. GOOD LUCK!

"We are what we repeatedly do,
excellence then is not an act, but a habit."

Aristotle
(Greek philosopher and scientist)

Cost Cutting Justification!

In any chemical industry and other industries, we cannot avoid from purchasing equipment or service which we need in order to perform our work to achieve our company objectives and targets. Whether we are working in a research institution or service company or chemical plant, we need to purchase something as asset(s) or consumable(s) or service(s).

As engineers/executive/managers, we need to make wise and technical decision based on wise justification before deciding to purchase something. We need to consider the economical aspect and the Return on Investment (ROI) as well.

However, from my humble experience, we usually know what we want and how the product/service works. We are also capable to evaluate and justify the product/service which best suits our demand and requirement.

However, the problem arises when our superior/procurement department questions us and tries to get the cheapest product/service without having better comprehension of what is exactly our requirement. Often, they'll overrule our demand and honour us with cheaper product/service that they manage to get. Their concept and philosophy is most of the time COST CUTTING! They do not want to compromise on the quality and safety aspects!

There is a well-known saying which states: *"Buy cheap, pay dear"*. An equally common expression is: *"If you pay peanuts, you get monkeys"*. The French say: *"On a rien sans rien."* (You cannot get something for nothing) while the Spanish equivalent is *"quien algo quiere, algo le cuesta"*.

You start to realise that we all pretty much agree the world over that if you cut corners on price there is a very good chance you will pay for it later. So why have we become obsessed with driving costs down irrespective of whether this actually has a detrimental effect upon us and our objectives?

If we have to discuss about this now, it will indeed take a long discussion. Obviously there are plenty of parameters that will decide certain decision. Sometimes, there are things that we don't know happening on the other side, or the administration side, or the accounting side. Decisions are made and there must be a reason for it. Bottom line is we need to be an excellent team player that works together with our top management. The best is to complement each other. Have some give and take. If what is needed for the plant to run is top priority and you require specific item, provide critical justification to the procurement department.

*"Engineers like to solve problems. If there are no problems
handily available, they will create their own problems."*

Scott Adams
(Creator of the Dilbert comic strip and the author of
several nonfiction works of satire, commentary, and business)

SCADA and Communications

People new to the industry often use the terms SCADA (Supervisory Control and Data Acquisition) and DCS (Distributed Control System) interchangeably. Although there's no sharp line separating these two categories of control systems, and specific implementations often have characteristics of each, it is useful to distinguish between them because of their unique strength and weaknesses.

The most significant factor differentiating SCADA system and DCS today is the "human in the loop". Enough about that explanation. It can be more technical and quite difficult to absorb for some people who are not really familiar with both control system. To make things simpler and easier to understand what are SCADA and DCS, let's see their characteristics:

SCADA System
1. Geographic span: Large
2. Point count: Large
3. Data acquisition rates: Moderate - seconds to minute
4. Data acquisition network: Slow with moderate error rate
5. Graphic User Interphase (GUI): Full featured
6. Alarming subsystem: Full featured
7. Control actions: Human initiated

DCS System
1. Geographic span: Small
2. Point count: Small
3. Data acquisition rates: Very fast - milliseconds to seconds
4. Data acquisition network: Fast with low error rate
5. Graphic User Interphase (GUI): Basic
6. Alarming subsystem: Basic
7. Control actions: Programmatically initiated

Perhaps, later I'll gradually explain about these types of control systems. We must know and appreciate these systems as it smoothen our process control and operation. These control systems are applied in the oil & gas as well as in refineries too serving the oils and fats industries, but not limited to those mentioned.

Heat Exchanger Cleaning & Maintenance - GASKETs

You can either totally dismantle the plate and frame heat exchanger OR you do the chemical cleaning. The decision you choose must be correct, as it will affect many things. You must decide which method to choose.

If you decide to dismantle the heat exchanger, this will give you the highest cost impact. Why? This is because, when you dismantle the plate & frame heat exchanger, the gaskets in between the plates cannot be used anymore. Once the heat exchanger is opened, the gaskets can't be used again. The polymer structure of the gasket has already been distorted and is not in a perfect condition to be used again. By the time you open the heat exchanger, some of the gasket, might also be damaged too. Some of them have become very thin. Some of them have been torn. There are many types of condition. And most of the time, oil leaks in the heat exchanger is due to the gasket problem or the operator/technician did not fix the gasket properly onto the plate heat exchanger.

One set of gasket to be placed in between two plates is very expensive. Depending on what type of heat exchanger, type of plates, type of oils/fluids flowing, and also range of operating temperature. This information can be obtained from a gasket or heat exchanger supplier/manufacturer. So, imagine if the plate and frame heat exchanger have 220 plates, you must have 221 gaskets placed in between the plates and the two stainless steel frames on both ends. On the other hand, chemical cleaning is cheaper, but may not solve your heat transfer problem if the scale and fouling is stubborn. The choice and decision is yours.

Centralized Control Room (CCR)

I was given a new responsibility by my superior few months ago. As a chemical engineer, I felt this task to be a very challenging but enriching one! I don't have any background on how this thing works! I was given the task of developing Centralized Control Room (CCR). I have to learn new stuff that I don't really know. I have to know all the new terms and meanings of all short phrases, abbreviations and wordings. I have to learn the control system of all the control system in all plants. It's a very interesting project to handle. There is not much physical object to be seen or touch such as cable/fibre optic/PC etc. I need to understand some network architecture in my plant and other plants at my workplace.

Well, there's something in Siemen product and learning centre about the CCR info. Siemen have a wide range of product for a very good and efficient way of control and planning in a plant/factory. You can visit and learn some stuff there. I'm also in the learning process too.

"Engineering is achieving function while avoiding failure."

Henry Petroski
(American engineer specializing in failure analysis,
a prolific author, a professor both of civil
engineering and history at Duke University)

Heat Exchanger Cleaning

You have a plate and frame heat exchanger in your plant. The heat exchanger has been in operation for perhaps one year already. By this time, the efficiency has dropped. The heat exchanger cannot provide you the desired output temperature when they heat exchange. How do you know this? You can know by seeing the output temperature of both fluids i.e. the cold fluid and the hot fluid outlet stream. If the output cold fluid temperature has increased and the output hot fluid temperature has decreased, then this is the sign of some sort of inefficient heat exchange situation. Why is this happening?

You can also check the pressure of the input hot or cold fluid, i.e. if the pressure increases; it is a sign of something called "fouling" which resulted in the heat exchange becoming inefficient. Yes, fouling is one thing that most of the time caused the heat exchanger to be not working optimally. All of these are just part of what you can notice and experience on your plate and frame heat exchanger. When this happens, you might want to consider removing the fouling from the plates.

But how? You have two options. As I mentioned earlier, you can both totally dismantle the plate and frame heat exchanger OR you do the chemical cleaning. Well, I can say the decision must be correct. You must decide which method to choose. Because both will have direct impact on your production and/or operation and/or maintenance cost. So, you don't want to be caught making a bad decision by your superior!

"A person's tongue can give you the taste of his heart"

Ibn Qayyim al-Jawziyya
(Medieval, Islamic Jurisconsult, Theologian, and Spiritual writer)

My first Heat Exchanger Encounter

It was in my third working place that I learn and have better comprehension about the heat exchanger. There are various types of heat exchanger in plant. We have plate and frame heat exchangers, shell and tube, spiral heat exchanger and others. The first time I got to know a plate and frame heat exchanger was when I was instructed by my boss to supervise a group of technicians and contract worker fixing and installing 233 plates onto a heat exchanger frame. It was not an easy job. The task was very physically demanding.

While performing this task, we have to be careful of the plate's arrangement. We have to be careful of the numbers and quantity of the plates. I have witnessed one time, these people accidently missed out one plate, which resulted in the 2 types of oil cannot pass through. This should not happen and can be a real upset. Imagine the cost and time involved. As soon as they realised that they missed out one plate, they decide to put another extra plate in between to continue the path of the two oils not to be blocked. Well, that's ok. But, they need to be careful next time. Not to be careless and repeat the same mistake.

The technique to tighten the plate and frame heat exchanger must also be correct. We must tighten it in a crisscross manner and at the same time take the measurements of the plate pack length.

There are still a lot more to tell about the hidden information in a heat exchanger. I mean the type of information and knowledge that is not written in the text book, the one that we can only learn from our own experience and other peoples experiences. I'll share some of my experience of dealing with heat exchanger from time to time. I sincerely hope the information will be beneficial for those who want to learn and know more about heat exchangers, heat transfer.

OFIC 2006 Exhibition

I learned and experienced a lot from attending Oil and Fats International Conference (OFIC 2006). The technical paper presentations were very interesting too. The companies participating in the exhibition came from various field and expertise which has significant impact on the oil and fats industry, particularly in the Asia Pacific region.

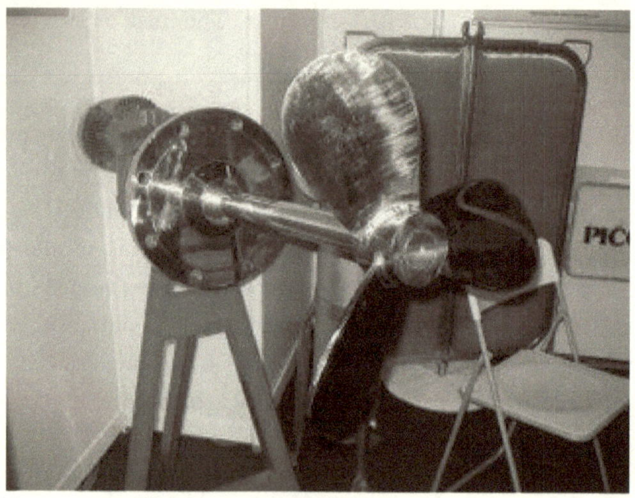

This is a side entry propeller for storage tank - to agitate crude oil or any product in a storage tank.

Some facts that maybe I can share with you: Malaysia currently accounts for 51% of world palm oil production and 62% of world export's, and therefore also for 8% and 22% of the world's total production and exports of oils and fats. As the biggest producer and exporter of palm oil and palm oil products, Malaysia has an important role to play in fulfilling the growing global need for oils and fats in general.

Various pressure leaf filters for refining oil industry. Learn more from the company supplying it - TapisTeknik

I managed to build relationships with the regions' multi-nationals and independent manufacturers/companies while interacting with them during OFIC 2006. Companies and manufacturers came from various backgrounds, disciplines and expertise such as plant & equipment, oleo chemicals, surfactants / detergents, specialty fats or instrumentation, bleaching earth, high pressure boiler, valves, motors, pumps, crushers etc. Participants and delegates came from Malaysia, Singapore, Thailand, Indonesia, Philippines, UK, US, Scotland, Belgium, Holland, German, India, Hong Kong.

A model of Lipico Biodiesel plant with 120,000 TPA capacity

Chemical Engineering Career You Can Expect

When I was doing my chemical engineering degree, we know that we are going to be a chemical engineer after graduating. For me, I don't really want to be a chemical engineer. Instead I would prefer becoming a chemical engineering lecturer. Most of my friends however, just want to be a chemical engineer. That time, we didn't really know in detail what does a chemical engineer do? If we know, perhaps we know it vaguely. Unless you've been to a practical training in related industry for a semester or more, then you'll know what I'm referring to. At least those students who've been in practical/industrial training are very lucky to have the glimpse and exposure of the challenging chemical engineering related industry.

As I mentioned in my earlier post, I've been in the chemical engineering research field, oil & gas/service engineer field and also plant/refinery field. I feel very lucky and blessed that I can taste various types of chemical engineering sector. Let me just brief a little bit about all those different sectors. Some might agree, some might feel, is that all? Some might, I hope, possibly get some valuable information from my sharing.

1. Research & Development / Post Graduate Study in a University.

This is the first thing I did after graduating. I became a research assistant for a very successful professor specializing in Reaction Engineering/Catalysis professor (Prof. Dr. Nor Aishah Saidina Amin (Famously identified as Prof. NASA)) specializing in Reaction Engineering/Catalysis. The research group is called: "Chemical Reaction Engineering Group" or CREG. CREG is based in Faculty of Chemical and Natural Resources Engineering, Universiti Teknologi Malaysia. When I first started, the group consists of 2 phD students and 2 master students besides several first degree students doing research related to natural gas to gasoline conversion and also palm oil to gasoline conversion using zeolite ZSM-5 catalysts. As I mentioned earlier, I didn't really know what the job nature is for a chemical engineer, and it includes the job as a research assistant. In general I know a research assistant will assist in doing research tasks. But, as days went by, I deeply understand the nature of research work. After about 6 months, I continued with my full time research Master in Chemical Engineering.

Let me share what research people do in general, point by point, not in particular order. The points are purely based on my experience with CREG. It may be useful for anyone who wants to do research related to catalysis and reaction engineering). Other research scopes in chemical engineering would not be the same like what I am going to share.

- Do research on the subject that you study in. Get technical papers from internet, library and other sources available at your university. One of the places where I get my technical papers is **www.sciencedirect.com**. Some, of the technical papers I got it from the subscribed databases in my university library. Some of them, I have to go to National University of Singapore (NUS) to get more technical books and references. I have a mass collection of this technical papers and books. I have both hard copy and soft copy; all which are very useful for my research.
- Create / construct your experimental rig. This is quite fun especially if you're that handy type of a person and creative.
- Synthesize catalysts. It took me 7-8 days to synthesize 1-2 grams of catalyst from scratch, based on the patent. Depends on what catalyst you are developing.
- Characterize catalysts. I learned using X-Ray Diffraction (XRD), Scanning Electron Microscopy (SEM), Fourier Transform Infra-Red (FTIR), Temperature Program Desorption (TPD), Atomic Adsorption Spectroscopy (AAS) and other characterization equipment.
  XRD: to study the crystallinity of the zeolite. We will know what type of zeolite it will belong to.
  SEM: to study the surface morphology of the zeolite catalyst.
  FTIR: to study what type of chemical bonding is available in the zeolite catalyst.
  TPD: to study the active acid sites on the zeolite catalyst by using ammonia gas. Hence, it is also always known as TPD-NH$_3$.
  AAS: to study the composition of metals available in the zeolite catalyst
- Run experiment at your rig. I did experienced running my experiment 24 hours a day, several days, sleeping in the lab.
- Publish technical papers, technical posters. Need to develop technical writing skills. Luckily I love this part. It will be tough if you don't like to write.
- Participate in technical exhibitions in and outside the country depending on your research budget.

- Present technical papers, technical posters which I enjoyed doing it too.
- Guide/supervise your juniors if there is. They can help you do some stuff too.
- Technical visit. This is fun too, you get to travel and learn.
- Undergo your first stage post graduate proposal. Hopefully I will pass this one.
- Prepare for your VIVA - Final exam. This is where you defend your research. This is a terrifying and suspense moment.
- Do research on the subject that you study in. Get technical papers from internet (as I mentioned earlier).
- Create / construct your experimental rig. This is fun especially if you're handy and creative. I constructed my own experimental rig which consists of a small 'carbolite' tubular furnace which has a maximum temperature of 900°C. I also have 2 digital mass flow controller fixed as well on the rig. It's used to control the flow of CH_4 and O_2 in the 1/2" quartz reactor. I also used several needle valves that we purchased from Swagelok.
- Run experiment at your rig. I did experienced running my experiment 24 hours. Normally it took 8 hours to run one experiment. After arranging the experiment according to the design of experiment (DOE) using response surface methodology (RSM), I have an arrangement of different flow rate of gases to control, different mass of zeolite catalyst and also different operating temperature. The duration of experiment differ from research to the other research.

2.   Service Engineer in an Oil & Gas field

After completing my master's degree, I was offered a job as a chemical application specialist/chemical technologist for a local oil and gas servicing company. But I call myself a chemical engineer. This category of work is full of fun because I travel a lot (If you like travelling, you'll like this one) I work on-shore and off-shore. I love it. One thing that I cannot forget is watching *"KILL BILL"* movie in the Dulang B Oil Platform (mother platform), in Offshore Terengganu. There are actually plenty of other precious experiences from this job which I will not forget.

It all began when a friend of mine asked, *"Hey Zaki! Do you have a job?"* She asked me that because I was completing my master's degree, and she guessed that I might want to search for a work. Well, I replied, *"Yeah! Of course I haven't got a job yet. In fact I'm still searching and applying for jobs via the internet."* Then, she told me about the offer. To cut the story short, after being interviewed by the company's top management, I secured the job.

Immediately after that, I was asked to join the team to go to Kerteh, Terengganu to handle a project. It was a new job in a new place and new working environment. I went to the site with a group of people. We had to attend a pre job meeting with the Petronas Gas Berhad (PGB) management in charge. In the pre job meeting, we discussed all about the project. We discussed the tentative program, schedule, working procedure, job safety analysis, emergency response procedure and much more.

After the meeting and after the PGB personnel are satisfied with our presentation and discussion, they'll approve the 'permit to work' (PTW). This is a very important document to have before performing any job at site. If we don't have this form completed and approved, we definitely cannot perform the job. It's against the PGB rules and regulation. And for other oil company, it's the same thing. When any accident occurs (we would not want this to happen), PTW is the first document to be referred to.

The project title that I am joining is basically *Internal Pipeline Chemical Cleaning* of the PGB pipeline. We were awarded a contract to maintain 2 pipelines (gas and condensate line). In short, we will pig the pipeline with various type of pigs. Well, pig is actually a tool to scrap and clean the inner side of the pipeline. Not the animal "pig". Besides pigging the pipeline, we add special chemicals to clean the pipeline and to prevent it from corrosion. This is actually my role. I was in charge in leading a group of people to blend the chemicals at site, do some in-situ testing to ensure the quality is according to specification, manage the process of injecting the chemicals into the pipeline etc. Not only that, there are more tasks and responsibilities behind that. This is part of the onshore story.

The *Internal Pipeline Chemical Cleaning* project is not a small one. Earlier, I mentioned that I took care of the chemical blending part. There's actually more than that.

Blending chemical is easy IF the chemical amount is small. But what if the chemical that you want to blend is huge? Yeah! I have to blend chemicals in big batch volume. When we performed a big chemical cleaning project, I normally will blend more than 25 m³. I blended the chemicals in a big 2.7m³ polytank. But not many of them were available. There were all together 5 polytanks only. So, I had to work and optimize the blending process using just 5 polytanks.

Here, the most important stuff is the PLANNING. I had to plan the blending process. I had to time the blending accordingly. I had to wait and collect water as a diluent for the chemicals. A lot of water was required. I used several submersible pumps as well in order to transfer water/chemicals from one polytank to another polytank. I used some explosive proof drum pump to suck out the chemicals from a 200L drum. I must prepare and make sure that power was available and it was supplied to our blending area in a very safe manner.

In this project I handled 5 types of chemicals/solvents.
a. *Degreaser* - to clean, dissolve any deposit and inorganic/organic material sticking in the internal pipeline. It's an alkaline water base chemical. It's a specialty chemical product produced by *ClearWater Chemical*. The type of chemical that we used is called Alpha-Clean W.

b. *Corrosion Inhibitor* - This is the chemical that will inhibit corrosion from taking place. It's also produced by *Clearwater Chemical*. We used a chemical called Alpha 3013. The appearance is dark blackish with pungent odour. If this chemical spilled onto your cotton coverall, don't bother to wash it because it won't go. And because of this chemical, we had to wear 3M half face mask with the acidic vapour and/or organic vapour 6003, 6006 cartridge.

c. *Biosolve* - This is the most pleasant multipurpose chemical. It is used to encapsulate hydrocarbon and also it can suppress any heavy metal vapour particulate in air. When there's any hydrocarbon spillage, we can quickly apply this chemical which will be first blended with water and just spray it on to the spillage. But beware. This chemical is expensive!!! But it's worth it. It's a great product. One application of *biosolve* is to suppress and neutralize hydrocarbon via bioremediation process.

d *Kerosene* - This is just normal kerosene that I use to perform some in-situ test to ensure the quality of the degreaser and corrosion inhibitor that I blended at site. I normally use kerosene to conduct the foaming and emulsion test. It's just a simple test.

e *Thiored* - this is a new chemical that I have to used and apply lately. This chemical is specially designed to contain and suppress mercury. It's a polythiocarbonate base chemical. We have to be extra cautious because there's a possibility that mercury in any form can exist in the condensate or gas products. And of course all these chemicals will be blended with water.

3. Process Engineer in a Refinery

It was very unfortunate that I cannot continue working in the oil and gas servicing company any longer. The management made a decision to move to Kuala Lumpur to get better business opportunities. I got myself a job as a process engineer in a plant processing edible oil. This is another chapter of a very challenging work. Here, I practiced most of my chemical engineering knowledge, specifically unit operation subjects. I admit that my learning curve here was the highest as each day was an adventure. Anything can happen. No day is the same day.

Boiler Problem

In October 2006, the utility boiler in my factory experienced some foaming problem. We knew foaming took place when we saw foam appeared in the water sample bottle. The water was taken from the autoblowdown line sampling point which takes out the surface boiling water inside the boiler. Although all the water parameters such as total dissolved solid (TDS), pH, alkalinity, phosphate, chloride content seemed to be within the control range the water chemical still create foam.

This is a great deal and matter because this can affect the steam demand and quality that will be consumed by all the plants in the factory. It can affect the production and cost hugely.

Although we have changed the boiler water, the foam was still there, but at least the foam will dissipate and disappear faster than before, showing sign of improvement. However, my concern and hope was to see zero foam appearing from the sample water. Well, we are still trying our best with the help of our boiler chemical supplier to eliminate the problem. I hoped the problem will be solved soon.

"Engineers are behind the cars we drive, the pills we pop
and the way we power our homes."

James Dyson
(Inventor, industrial design engineer & founder of the Dyson)

Terrifying Moment at Boiler House

Few days ago, I posted something about the condition in the boiler at the plant that I work in. We tried to solve the problem and after consulting with our water treatment boiler chemical vendor and his technical expert in America, it was decided to do a caustic boiling inside the boiler to remove all possible traces of oil which was suspected to be inside the boiler that causes the foaming issue. We added up caustic, trisodium phosphate and sodium phosphate and heated up to about 60°C. After monitoring the chemical blending process and pouring the chemicals into the top manhole of the boiler, I left the boiler and did some other jobs. Later in the afternoon, after visiting the control room in my plant, I paid a visit to the boiler to see the progress there. I tried to ask my colleague (the person in charge of the boiler), but he was busy doing something on the boiler control panel. At that moment, I was standing nearby and diagonal to the boiler. Suddenly, there was a very loud explosion.

I can slightly see some red colour flame on my right. The incident happened very fast. I was in total shock and immediately tried to brace down to cover myself. Then I thought, I should seek cover to protect myself by hiding behind the small boiler house/room just on my left. It was difficult to open my eyes because of the smokes and thousands of fine particles/dust which were covering/flying around the area. I fear there might be a greater secondary explosion that might wipe off the boiler and surrounding area. If that happened, we might be history! When things became OK about 30 seconds to a minute later, I have a clear picture of what actually occurred. I can see smokes and some people started gathering closer to check out the incident. The burner of the boiler which has few dozens of strong bolts holding them was lying on the ground next to the front of the boiler. Thick black coloured oil was expanding and flowing slowly from the burner. All the piping and other materials nearby the burner was damaged. It was not a nice scene.

Then I saw some people coming down from top of the boiler. I was surprised that there were people there at that time. My colleague, the steam engineer, the chemical vendor and few maintenance fitter came down from the boiler and were surprised to see the burner was ripped off from the boiler body. At that moment, we were wondering, what caused the small explosion to occur? Investigation and theory on what went wrong began.

"Our virtues and our failings are inseparable, like force and matter. When they separate, man is no more."

Nikola Tesla
(Inventor, electrical engineer, mechanical
engineer, physicist, and futurist)

Boiler Explosion: What Went Wrong?

After almost one week since the small boiler explosion occurred, I asked my colleague, the person in charge of the boiler, what was actually the reason the boiler exploded? I asked him just now, because the complete report from the specialist has been revealed.

Ok, we had some problem of some slight oil contamination inside the boiler water/steam. This is definitely not good. It can create some fluctuations in the steam production and also can cause wet steam which will result in vacuum pressure drop. And there are a lot more reasons on the bad effect of oil being inside the boiler.

So, the chemical vendor suggested us to do the Caustic Boiling. It's a method to remove completely any traces of oil inside the boiler. So, we did it. The chemical is supposed to be mixed and boiled inside the boiler at temperature ranging from 60 - 70°C. The feed water temperature is already about 80°C. Our superior advised to heat up the temperature so that the caustic can be thoroughly mixed with water inside the boiler via physical boiling technique. And guess what, the man-hole was not closed.

As the temperature increased gradually, the water + chemical are now almost boiled, and the water splashed out of the boiler man-hole. The hot water + hot mixture of various caustic came out of the boiler man-hole and flowed down the external boiler body touching and covering the boiler control panel. Luckily my colleague bravely and swiftly shut off the boiler power/panel despite knowing that he can badly injure himself if the hot caustic fell and splashed on his body!!! Immediately the electrical and instrument departments were called to save the control panel by drying it to avoid wire/electrical shock later.

Later, the electrical and instrumentation department worked hard to dry everything on the control panel. The paint on top of the control panel was damaged and ripped off from the effect of hot caustic. The boilerman also worked hard to clean the mess and spillage in the surrounding area. The boiler was left for a while to let it cool down and dry.

After about 2-3 hours, I guessed, instruction was given to start the boiler, at low fire mode, just to slightly heat up the mixture of caustic and water in the boiler. The time was about 1430 hours, and I just arrived at the boiler and was watching the boiler people doing their job. That was the time when the boiler exploded and produced such a terrifying sound!

After investigation, it was found that burner A (it's a twin burner boiler), furnace backfired. In any normal operation, during initial start-up, combustible matters must be evacuated from the boiler by pre-purging. This can be done by using the boiler control panel. The boilerman intended to run the boiler as usual, but instead, without anyone realising, the boiler burner sequence at the control panel malfunction. It is believed that the boiler which was already contained with natural gas, unexpectedly switched into the high fire mode, which triggered a huge spark. That caused the explosion to occur which made the burner ripped off from the boiler body and created great damage around it. The boiler program sequence actually went wrong, because of the short circuit in the control relay base from the earlier hot caustic spillage

It costs a lot to repair the boiler. I solute our maintenance fitters, instrumentation and electrical technician, including the boilerman because they worked very hard to rectify and repair all the damage. Within a short period of time, about 36 hours later, the burner (which can luckily still be used) was re-attached to the boiler and all pipings, fittings, valves, gear etc. was fixed back. We called external boiler experts to help us with the burner sequence and start-up operation, and that cost some money too. Now the boiler is back in operation. And for sure, we'll be extra careful from this moment to avoid similar or almost similar incident from occurring again. I will never forget that moment. And another moral of the story, do not walk unnecessarily nearby the boiler! Appreciate life.

Centralized Control Room Dilemma

We were planning to have our centralize control room (CCR) in our group/factory. We planned to include SCADA and Manufacturing Execution System (MES) in the picture. However, after some discussion with my superior, it seemed quite complicated to incorporate SCADA and MES in the CCR. If we want to include MES, that will include production scheduling etc. and a lot more departments such as operation, production, shipping etc. We think that is quite difficult to handle via MES because a lot of thinking is involved in a lot of critical daily decision making.

Therefore, I told my boss that we are actually not looking at MES and not SCADA as a whole, because SCADA will include some extent of human control via the human machine interface (HMI). What we want in the CCR is just some monitoring of the vital plant parameters, some analysis on the Overall Equipment Efficiency (OEE), Overall Plant Efficiency (OPE), things like downtime tracking, Key performance index (KPI), and probably some other stuff that I need to think of to make the CCR a really useful, easy and imperative tool for the top management to use. I must select the most vital plant tagging as well, check and compare it with the list of digital input, digital output, analogue input and analogue output at the control panel. Gosh! This is a very tough assignment for me!

*"The proper method for inquiring after the properties of
things is to deduce them from experiments."*

Isaac Newton
(Mathematician, Astronomer, Theologian, Author and Physicist)

Chemical Engineers Habit

I don't know about other graduate engineers or chemical engineers, but for me, as engineers we need to have our target in career and profession. We need to plan everything since the very beginning. We need to know our destiny. But, sometimes, it's difficult. Sometimes planning seems to be very easy, but the outcome might not be what we want.

Continuous learning is imperative for every engineer. For me, I appreciate every knowledge around me. I love knowing and learning new things. I set my own objective, or practice, which is **to learn new things every day**. That's why everywhere I go, I'll bring along something to read. It doesn't matter whether the reading material is a book, magazine, printed paper or scribble of paper which I can read or develop ideas from it. I subscribe to a lot of free technical magazines in order to get new and fresh information. There're a lot of those technical magazines that are available for me to read in my car as well. Among the free technical magazines are those that I have listed at the Recommended Resources at the end of this book. Some of the magazine I subscribed online. That'll be easier too. As I checked and browse through my email, I can read them too.

I'm happy and appreciate the pleasure to learn new knowledge every day. I can post many things on my blog daily, but it's the time factor that prohibits me from doing so. Therefore, I only post something that I think will benefit my guest (readers) and that's part of my intention, which is to share chemical engineering knowledge and experiences. If a chemical engineer does not have the desire or fire or excitement to learn and learn and continue learning...that's not a good sign! For me he/she cannot develop and become a good knowledgeable chemical engineer.

My final say on this topic, ensure you learn new knowledge/ideas daily. Enrich yourself. Set your target. I set my target too. And my back-up target as well. Because life is not easy, sometimes we need plan B, plan C, and plan D and so on. Because there are always chances that plan A would not work!

I'm not saying that I'm a good engineer. But I want to become one. Let's do it and work it out together.

Sudden Pigging Job

In May 2006, I called a friend of mine whom I worked with in my previous company (the oil and gas servicing company), to have a chat We talked a little bit about some news and info about the jobs, projects and company. He suddenly asked for my help. I asked what is it? He asked whether I'm free next week, between Wednesday till the weekend. I asked him what actually he wants me to do. Then he explained that he wanted me to join him and his team to go to Songkla, Thailand to do the pig retrieval job there. I was surprised when he asked that sort of favour from me.

The 36" filming pig was launched from Cakerawala Platform at the border of Malaysia and Thailand about 1 week ago. It was COSS first trial project there and their first attempt to go international. I worked with COSS previously for about 2 and a half years and I handled a lot of pigging jobs with them, but I have not handled any international pigging job yet. This might be an interesting and exciting task to participate.

The problem that my colleague has is he has lack manpower that has the international passport and expertise to perform and manage the job immediately. That's why, when I called him, he immediately asked for my favour. Though I'm dying inside to go and venture new experience in a new place, I can't go. I have a lot of works, projects and a plant to take care of and handle in my existing company that I work now. It's a blow for me. And I know my boss would never let me take any annual leave that long. I just wish that I can have a more flexible job that I can freely perform any other technical jobs/projects if invited without affecting my present job. Would that be wonderful? I can further develop my overall skills and career.

Cooling Tower Water Treatment

Taking care of the cooling tower water is not as easy as it looks. When cooling tower water looks crystal clear, that means whoever is taking care of the cooling tower is doing a very good job. It also means that the process plant people are also doing a great job as a customer of the cooling tower water receiver. It also means that the operational part of the cooling tower is excellent. The mechanical part of the cooling tower is also in a tip top condition shape. The heat exchanger should also be performing to expectation.

You can expect some darkish slimy film sticking on your infill when oil leaks. Well, basically what a cooling tower water needs for treatment is as follows:

1. Biocide
2. Corrosion / scale inhibitor
3. Oil dispersant

The specialty chemical required for any cooling tower depends on the cooling tower situation. What I've put earlier is just the basic stuff. We must dose the right chemical at the right time, if not, the consequences might be devastating.

Well, you can Google and read more about one company, Nalco, that have a cooling tower water chemical treatment. Our cooling tower use chemicals from Nalco. However, there are other brands available in the market for your cooling tower water treatment too.

Plant Performance - Heat Exchangers Dilemma

I have been looking and concentrating to improve my plant performance. Sometimes, it's difficult to get the desired oil quality and the management will question this. Sometimes, I cannot produce the target throughput that I was supposed to deliver. Some of the prime reason is just because of inefficient heating in the plant, where one of the reasons being my heat exchangers are not working effectively.

Therefore, I cannot get the heat that I wanted which directly affects the oil product quality. Now, my major task (together with the others) was improving all the heat exchangers so that they are properly maintained. I need to have extra fresh plates for plate heat exchanger so that we can immediately have it for back up. I also need to add some more "cleaning in progress" line, to clean the heat exchanger on-line. This is among the challenges that I have to face.

In the meantime, I have to work with some of the less effective heat exchangers. If the heat exchangers are not ok, the plant will consume more steam and natural gas/fuel apart from not getting the quality. Anybody facing the same problem....?

"The engineer has been, and is, a maker of history."

James Kip Finch,
(American engineer and educator)

Schmidt-Bretten Plate Heat Exchanger

I used a lot of plate heat exchangers (PHE) in my plant and most of them are manufactured by Schmidt-Bretten. The model/type of plate heat exchanger that I used is Sigma 64. There are other brands and types of PHE, but this Sigma 64 is the most available in my plant. Just like the one that you can see in the next page, but that one is fouling badly. You can see the pieces of hard gum and scale etc. Ugly stuff!

I'm having a tough time gathering information and inventory of my spare heat exchanger plates. I need to back up my plates so that I can dismantle a heat exchanger and replace them immediately with the spare plates that come with gasket. This is very important to ensure production smoothness. There's a lot to tell about a small topic of heat exchanger. We do not get this in university and text book. We get most of this from experience.

*"If you want to succeed, you should strike out on new paths
rather than travel the worn paths of accepted success."*

John D. Rockefeller, Jr.
(American financier and philanthropist)

Dismantling Plate Heat Exchanger

Following are the photos of plate heat exchanger (PHE) that we recently decided to open up/dismantle because we cannot get good heat exchange and the desired temperature. The PHE has been in operation for about 2 years already. We can see gum from crude palm oil (CPO) sticking on the outer part of the plates which created fouling on the PHE. All together there are 115 plates on this PHE. Now, we are dismantling it too clean the plates.

See how bad is the fouling on this PHE. The black hardened gum from CPO was found sticking on the plates. Some of the pieces of hardened gum have been removed, and we can slightly see the plates. Can you see it?

Vacuum System Failure Rejection

Another thing that I don't like while running a plant is having a down time. I hated it most when oil rejection occurs. Rejection will happen for various reasons. One of them which occasionally happens is the pressure vacuum drop. Oohh... when this happen, the oil free fatty acid (FFA) content or oil colour will be out of spec. This will force my plant operators to reject the oil to their respective crude oil tank. I'm still learning and exploring as well about the vacuum system in my plant.

I need to maintain the dirty cooling tower pressure and temperature within the desired range. If the cooling water temperature slightly increases, it can affect the vacuum system and resulted in vacuum drop. It's ok if I know and can explain why the down time or rejection happens. I hate it when I don't know the reason. It makes my life a little bit difficult explaining and answering all sorts of questions bombarded at me just because I cannot meet the daily production target. Well, this makes me want to really make sure that all the processing parameters are within range and control. Need to prevent any plant breakdown and downtime from happening.

"Problem-solving is essential to engineering. Engineers are constantly on the lookout for a better way to do things."

Dinesh Paliwal
(President and CEO of Harman
International Industries)

Throughput versus Quality - Heat Exchanger Dilemma

I'm in a deep headache at this time. I have to deliver a throughput of 3000 MT/day. Now, one of the twin plants was having a small problem. My plate heat exchanger was suddenly not giving the sort of heat transfer that I wanted. As a result the quality which will be achieved by having high temperature could not be obtained. I have to lower the flow rate of oil passing through the heat exchanger so that better heat exchange can take place.

The heat exchanger can cope with 40 m³/hr flow rate that I normally run, but the temperature of 230 could not be achieved. So, I have to lower it down to about 34 - 36 m³/hr. Around this time, the shipment was very tight which means I have to deliver processed oil which always must be above 3000 MT/day. The heat exchanger must be stopped and cleaned. I must replace it with another spare heat exchanger. The problem is my spare heat exchanger is also having some other problems. Ohh goshhh!!!

"I've always been an engineer devoted to the potential of advanced technologies. Like most engineers, I have a keen sense of curiosity and a deep desire to learn. Garmin was my first entrepreneurial endeavour, and it has been an incredible journey."

Min Kao
(Electrical engineer, businessman and philanthropist,
co-founder of Garmin Corporation)

Taking Care of a Physical Refinery Plant

The pass 2 weeks was a very tough one for me at work. And it's not going to end. This is just the beginning. A major part of my job is to handle a very big physical refining plant, with a capacity of 3000MT/day running 24 hours for about 355 days a year together with a very experienced senior production executive. Another major part of my job is to take care of quite a number of projects and utility jobs in the factory. This consumes a lot of my time and as a result, my duty and time for plant is not much. I always depend on my senior colleague when it comes to the plant matter. I just started co-taking care of the plant with him for about 1.5 years now.

My worst nightmare has come. My senior colleague was offered a very good job and position elsewhere. The payment is awesome too. For that, I guess, I'm going to take care the plant alone. I really don't mind taking care of the plant if I have all the knowledge and skills to control and take care of the plant. The problem is that the knowledge and skills are still insufficient for me to take care of the biggest physical refining plant in south Johor alone. I need some assistance. Between last week and this coming week, I'm racing against time to learn as much as possible everything in detail about the plant from the senior production executive. It's a very big plant. It's a very fast plant. It's a very important plant. I cannot afford to make any mistake in delivering instruction to my subordinates or controlling the plant. I have to be 24 hours alert of what's happening to the plant because the plant is running 24 hours.

This job might look simple for those who have the experience. Unfortunately for me, I'm still considered new and fresh to this plant although I know and can understand the theoretical part of the plant. Well, I have to face this challenge. I have no choice. There're no other people to fit in here. I'm going to face it. Wish me luck you guys. Pray for me. God help me please.

Not A Chemical Engineer, But Performs Like One!

I worked in a physical refining plant and currently take care of a big plant. Before this, the plant is always taken care of by two executives/engineers. Now, my senior colleague (we took care of the plant together), Mr J is going to leave because he was awarded an excellent job offer from an international competitor company. The payment is a lot too. I'm not really sure, but my guess is that he is going to be paid 4 times of what he is earning from this company we work. What I want to highlight here is everything about him. And I'm going to take care of the plant alone now.

All this while he has been my senior, mentor, guru, colleague, etc....you name it..... He is very good in almost every technical aspect in the edible oil physical refinery and industry. He managed to take a very good care of the plant and has help improved the plant production capacity/upgrading from less than 2000 MT/day to about more than 3000 MT/day. He has been working in the present factory/plant for almost 13 years and in the industry for almost 30 years.

What impressed me most is his educational background. He is not a graduate. He does not have any certificate. His only education is the high school/secondary school. He started to work in a factory not as an operator but as a cleaner (this is what I was told, I'm sorry if I'm wrong). During work, he saw plant operators operating the plant and became interested. He quietly learned and he became a plant operator. He was diligent and he was promoted to become shift leader and later supervisor. Soon, he was offered a better and more challenging job in the place where I'm working now as a production executive. Later he forced himself to become a senior production executive. And now, he is going to be a very important person in the new company that he will work for.

How can he become like that? It's the attitude, hard work, dedication, determination and most importantly the willingness to keep learning and learning. Although he was only a high school leaver, but he managed to become better than a practicing chemical engineer. He forced himself to converse in English. He did mention to me he took an intensive course to learn English. He practiced and learned computer skills to do reports etc. He learns management skills etc. He learns all the necessary technical stuff. He learned about pressure, temperature, flow rate, vessels, boilers, cooling towers, piping, PLC, control system, NPSH and hell lot more...all from his exposures and experiences. Huhh.....what a fantastic person he is.... We, engineers should have better or at least similar attitude and work ethics like what he has. Period.

"Save future time by planning and listing down what you are going to do tomorrow. Then execute the list the next day"

Zaki Yamani Zakaria
(Founder of Chemical Engineering World Blog &
Chemical Engineering World FB Page)

Hot Water System Problem

Few weeks ago, our specialty chemical vendor from Nalco came and performed his routine water testing in one of the plant in my factory. He was testing the hot water tank system and found out that the water pH was below 6. The pH is supposed to be controlled at about 8. Other chemical parameters seemed OK and within normal range. So, he added some caustic solution into the hot water system to increase the pH. He also advised us to dispose half of the water in the system and introduce fresh water. He gave me a copy of the report and explained to me the situation.

Later, the same day, a supervisor from the plant was instructed to flush/blowdown the water in the hot water system. As soon as he opened the final valve, hot red coloured water came out from the system. We were surprised with the colour of the water! My boss who saw the colour of the water immediately thought a severe corrosion is taking place inside the system. I went to the site observed, took some photos and took some sample as well.

From my previous experience, I know that, if corrosion really takes place, the iron metal which is heavier will settle at the bottom of the sample bottle. I waited for one day and the next day, I found out that the water is still in the same state and colour. No changes can be observed from the colour. This is a very homogeneous solution and I believed that this is not corrosion. It was something else.

I waited for the test result which the chemical vendor carried out. Later, the following day, he emailed me the report and the result showed that iron is undetected! First conclusion ---> No corrosion is taking place.

Later we realised that the red colour water is actually a product of a chemical reaction in the hot water system. When the chemical vendor poured in some caustic solution to increase the pH, the solution actually reacted with some oil inside it. We are still not sure which type of oil (Lab test indication). Second conclusion ---> one or more heat exchanger is leaking and therefore, the plant personnel must detect which heat exchanger is leaking and isolate them and rectify the leaking heat exchanger.

The physical condition of the hot water system

"Science is about knowing; engineering is about doing."

Henry Petroski
(American engineer specializing in failure analysis,
a prolific author, a professor both of civil
engineering and history at Duke University)

Flood Indirect Impact

These were among the reasons why we were having CPO shortage! This happened in Segamat (refer image below). In Kota Tinggi, about the same thing happens. The CPO tanker was already trapped in the fierce flood. As an indirect result from this, my plant was forced to have a shutdown on 31st December 2006, Sunday. Hmmm and guess what? That was a public holiday here in Malaysia because of the Haj Celebration and also the next day was the New Year!!! I had to work on that public holiday!

A CPO tanker stuck in the flood.

Flood attacked my country in 2006

Heavy Rain & Flood Impact to the Industry

It was still raining. The rain kept on pouring, heavily. The state/country was hit with one of the worse flood in history. Luckily my house and the surrounding area were not flooding. However, it gave a significant impact to my work. Not just when I went or returned back from work, water clogging (minor flood), very bad traffic jam, a lots of holes on the road, some damage to vehicles etc.; it also forced us (my company) to be prepared for a plant shutdown, simply because of the crude palm oil (CPO) supply shortage.

Flood also impacted the normal business activity

Another scene of the flood

We never expected such scenario to occur. Due to heavy rain, a lot of palm oil plantations were badly affected by the flood. Hundreds of workers associated to palm oil harvesting and milling could not go to work. CPO Tanker could not reach us because the road structure collapsed. We have a serious shortage of CPO. Most of our CPO tanks in the factory were empty. If the CPO runs out, we have to stop the plant! We cannot blindly and easily stop the plant just like that. Therefore, we had to take the opportunity to perform any maintenance work. That's why I was quite busy alongside with my colleague preparing for the shutdown. I also have to carry the umbrella everywhere.

Not just that, the CPO price has already escalated to nearly about RM2000 per metric tonne (that's about US$570/tonne). I could not recall the price rise up that high in the past.

"Strive for perfection in everything you do. Take the best that exists and make it better. When it does not exist, design it."

Sir Henry Royce
(English engineer and car designer)

Visitors to Plant

Last week, my scheduled was so tight. In a space of 2 days, I have to standby and wait for three different visitors who happen to be very important to the company. I have to ensure the plant is in clean and in good condition. Although the day and date is known, the exact time of visit to the control room and plant is unknown. There was no indication of what time they were coming. Therefore, I have to wait and do whatever work I can do in the control room. Well, that's fine; at least I can do a little bit of work using the PC available in the control room.

The first group of visitors were from Australia and they were oil traders. The second group was from the Iran ministry of health. The third group was from the ministry of Sabah. As for me, when all these visitors came, I have to explain briefly about our process on how we convert crude palm (CPO) oil to refined bleached deodorized palm oil (RBDPO). The explanation was made easy because I can use the SCADA system in the control room to present briefly about our process.

"Genius is patience."

Isaac Newton
(Mathematician, Astronomer, Theologian, Author and Physicist)

Some Comparisons of My Jobs

Since I graduated in 1999, I have been in three different chemical engineering fields. The first is Reaction Engineering, Chemical Engineering Research (Aug 1999 - Dec 2002) which I gained my Master's Degree from. The second is *Chemical Engineering in Oil and Gas Industry* (Jan 2003 - June 2005) and at present, CHEMICAL ENGINEERING IN OIL AND FATS INDUSTRY (June 2005 – present (2007)).

If asked...which one I enjoyed the most:
CHEMICAL ENGINEERING IN OIL AND GAS INDUSTRY - I tasted and experienced a lot of interesting stuff

If asked...which one I learned a lot (really practicing chemical engineering fundamental):
CHEMICAL ENGINEERING IN OIL AND FATS INDUSTRY - I practiced a lot of Unit Operations

If asked...which one is more relaxing:
Chemical Engineering in Oil and Gas Industry - Maybe because it's a service company.

If asked...which one is more challenging:
Chemical Engineering in Oil and Fats Industry &
CHEMICAL ENGINEERING IN OIL AND GAS INDUSTRY - Both have different angles of challenge.

If asked...which one I prefer to be in now:
Chemical Engineering Research - I love doing research + publication.
If asked...which one I make the most money?
CHEMICAL ENGINEERING IN OIL AND GAS INDUSTRY - Undeniably the wealthiest field.

If asked...which one really consumed a lot of my time:
Chemical Engineering in Oil and Fats Industry - I just work, work, work till late....?

If asked...which one I have more down line manpower:
CHEMICAL ENGINEERING IN OIL AND FATS INDUSTRY - I have
skilful operators to help running the plant

If asked...which one I felt more pride:
CHEMICAL ENGINEERING IN OIL AND GAS INDUSTRY - Well, it's
a proud feeling being in this industry

If asked...which one allows me to travel more:
CHEMICAL ENGINEERING IN OIL AND GAS INDUSTRY - Went
overseas as well as offshore...from a four wheel drive Pajero to a Sikhorsky 76
chopper....from an antic Mercedes to a Boeing 737 aircraft....from
Singaporean dollar to Rupiah.

If asked...which one allows me to do more publications:
Chemical Engineering Research - I published quite a number of papers and
manuscripts with my supervisor.

If asked...which one gives me more exposure:
Chemical Engineering in Oil and Gas Industry - Living in an oil rig platform is fun,
interesting and makes me eat all the time, provided if you don't stay there 4
weeks in a row.

The above points are just my opinion and purely what I feel from my
experiences. For your information, I did my full time research in **University
Teknologi Malaysia.** I worked in a **service company** which deal in oil and gas
as a chemical engineer cum project engineer. Latest, I am a process engineer
taking care of a physical refinery plant processing mostly palm oil and
sometimes some lauric oil.

Found Abnormality, Take Immediate Action

Just now in the afternoon, I walked pass my plant and I saw the suction pipe to the 37.5kW pump delivering water to one of my cooling tower was vibrating badly. I have a closer look to inspect it. I saw water dripping from it and I felt that the pipe was already very thin. The pipe has experienced bad erosion from the acidic water flowing through it. I immediately called my plant operator and asked them to switch pump, and called maintenance team to replace piece of the pipe which is already beginning to break down into pieces. After issuing a job order to rectify the situation, maintenance team prepared their stuff and immediately come to the site and replaced the piece of pipe which was badly fractured. Within 2 hours from the time I informed my operator, everything was sorted out smoothly and the suction pipe to the pump is ready to resume operation.

Morals:

1. When you see abnormality(s), immediately report. Take action.
2. Anything that might affect your plan process/system must be immediately rectified.
3. Never delay any abnormality or problem - you might forget about it, and after a few hours of not taking action, it might be too late!
4. Follow up! Do not take any instruction you gave for granted. Maybe your down line delays it. Create urgency in the instruction depending on the situation.
5. I cannot think for some more....I'm exhausted....

Understanding Plant Tips

For you chemical engineers out there who happen to become a production executive or a process engineer who has the responsibility of taking care of the plant, here are some useful tips to ensure you understand the plant as fast possible.

1. **Learn all the piping system in your plant.** With this you can have better comprehension on the plant process.

2. **Learn how to start and stop the plant.** This is always the important thing to know. If the plant is having problem - you stop it. If problem solved - start it back.

3. **Don't be embarrassed to ask the supervisors or plant operators about the plant.** In most places, they have been operating the plant for a lot of years. They have tremendous experience on operating the plant and some trouble shooting. You can ask about the plant and asked from a lot of them. Therefore, you'll get more information from a lot of operators, and most of the time the info that you need will be common.

I hope those tips can be useful. It is a continuous process of learning. Furthermore, the plant that I'm handling is very big (size and capacity!). I would like to thank my senior colleagues (2 of them) who gave me those useful tips. I appreciate their kindness of passing me the knowledge, information, skills and tips. Thanks as well to my supervisors and plant operators for their hard work and duty in running the plant. Not to mention all the maintenance, electrical, instrumentation team. I hope my plant will continue to run smoothly...

Overall Heat Transfer Coefficient (U-Value)

I have been collecting data for the plate and frame (PHE) heat exchangers in my plant. There were 7 PHEs that I closely monitor. There were several more that I do not really focus. Among the data collected were inlet & outlet temperatures, flow rate and surface area. My aim was to calculate the "Overall Heat Transfer Coefficient (U-Value)" so I can know which PHE has become heat transfer inefficient. I created and formulated an excel file to make it easy for the plant operators to key in the data.

After a series of data gathering, I observed that sometimes the PHE generated high U-value and sometimes the same PHE shows lower U-value. I make a conclusion that the U-value in my case is actually strongly influenced by the flow rate. In my plant, sometimes we have to increase or lower down the flow rate mainly because of quality control issue. In addition, from my point of view, the effect of Log Mean Temperature Difference (LMTD) is not as great as the flow rate effect.

Later, I embedded the formulation in our SCADA interface in the control room. From there, the U-Value will automatically be generated and it is easier to monitor which PHE need to be checked for cleaning or plate replacement consideration.

"The future belongs not to those who merely seek opportunity, but to those who create it. Let ushave the courage to do things differently"

Dr. K.Anji Reddy
(Ph.D Chemical Engineering from National Chemical Laboratory,
Pune, Founder of Dr. Reddy's Laboratories)

Plant Process and Oil Lost

While processing *crude coconut oil* (CNO) to *refined bleached deodorized coconut oil* (RBDCNO) early February recently, I faced one problem. The problem was that the temperature control was slightly on the higher side. Normally we should control the deodorization temperature between 240 - 245°C. However, the temperature went beyond that, at about 246-249°C. As a result, in the pack column, where we are supposed to strip off *coconut fatty acid distillate* (CFAD) via vacuum, some of the RBDCNO escaped together with the CFAD. Later, we realized that our CFAD tonnage was twice higher than what we normally achieved. That is not supposed to happen, but it does reflect that we were losing some RBDCNO unnecessarily. I can be in serious trouble!

The price of RBDCNO is far more expensive than CFAD. Assuming we lost 20 tonnes of RBDCNO which flowed and mixed together with CFAD, and assuming that the cost difference between RBDCNO and CFAD is about RM1500/MT; therefore 20 tonnes x RM1500/MT = RM20,000.00 lost worth of RBDCNO resulted from this. Sadly speaking, that happened just because the temperature control was not within the designated range. Coconut oil easily breaks off at temperature's more than 246°C, unlike palm oil which can withstand temperature up to 270°C.

Morale of the story, some guidance or working instruction was prepared to be followed. During plant operation, we should follow the designated process parameters so that we can achieve our target quality and production. Despite the ISO9001 instructions available, all of us overlooked this temperature control issue. My bad...

Checking the Valves

Before facing the shutdown in 11 days' time, I toured my plant to check all the valves and other parts that need to be replaced. I went from one pump to another pump. There were a lot of pumps. They don't look good. Ugly black dirty oils covering them. I checked the suction and discharge ball valves. Some of the suction valves are 6" and 8" and the discharge valve are 4". Well, the sizes of big valves are easier to detect or decide. There's no big problem for me in identifying them now. However, that was not the case when I first joined the company earlier. It took me some time to get familiar with all types of valves, their brand and purpose.

Then I went from one heat exchanger to another heat exchanger. I checked out the drain valves. It's harder for me to estimate the sizes of those drain valves and deciding the brand. They are commonly small varying from 1/2" to 2". In between those sizes, there are also 1.75" 1", 1.5" and others. Deciding the correct size and brand is vital because we have to order and purchase the repair kits/seats. Ordering the wrong spare parts will just be a total waste of money and time and the chance to replace the valve seats which can only be done during plant shutdown. The darkish, oil layer covering the valves at the pumps and heat exchangers are something that must be improved. It should not be left dirty and uncomfortable for eyes. It should be cleaned and well maintained. I want the valves to be clean and shine like a new piece of metal. We are working on that as well. I implemented the 5S system (a Japanese Concept of Total Productive Maintenance (TPM)) for maintaining the plant to be a clean healthy place and pleasant working environment.

Checking Valve Repair Kit

The plant shutdown was just around the corner. Early in the morning, I went to the store to check on some ball valve repair kit for the shutdown maintenance job. The repair kit is mostly the internal part of the valve that can be replaced. The valve seat need to be replaced because of simple reasons such as it has already been worn out which resulted to leakage and also because of other reasons like the parts are jammed etc. There can be other reasons that could lead to the valve seat replacement.

In my plant, most of the valve brands used are Worcester and Valtac. Therefore, I checked most of the Worcester and Valtac repair kits which varies from 3/4" to 6". The valve repair kit is also not cheap. They are quite expensive especially the black seat type which can tolerate high temperature. The oil flowing inside my refinery plant may reach up to a maximum of 270ºC. That's why we need a good quality repair kit that can withstand great temperature and pressure.

Actual Working Hours

My official working time as a *process engineer* is from 8.30 am to 5.30 pm with lunch break interval from 1.00 to 2.00 pm. But, I normally go home between 6.00 - 7.00 pm and hopefully reach home within an hour or more. The distance between my home and work place is 50km (I commute 100km back and forth from work).

Within this limited space of time, I will try and do as much work as I can. However, there are two activities that further minimize and reduce my working time. The first one is "meeting" and the second is "interacting with suppliers". Let me share an example.

Every day, I have a minimum of 2 meetings which are held from 9.30 - 10.00 am (Maintenance/Project Meeting) and 11.30 am - 1.00 pm (Production/Coordination Meeting). Total time spent for those meetings = 2 hours. That's why sometimes when my wife called, I have to say: *"I'm in a meeting, I'll call you back later, honey..."* xoxo

Upon reaching 1.00 pm, I'll have a break. This is my free time where I'll spend it either for lunch or short nap or surfing for information in the internet. Sometimes, I did some internet surfing.

After lunch break, a supplier supplying filter sock came and met me. We talked and discussed few things regarding the filter sock problem and solution. Frankly speaking, I don't really know about filter sock. Therefore, anything that came out from the supplier's mouth, I just absorb it.

Later the same day, another supplier selling bleaching earth came and begged us to purchase bleaching earth from him. We have not been ordering his bleaching earth since last November and he claimed that his business is getting bad to worse. We have our reasons for not ordering the bleaching earth from his company.

Meeting those two suppliers already cost me another 2 hours where I resumed working at 4.00 pm. Let me just simplify the situation:

Total official working hour = 8 hours

Time spent for meeting = 2 hours
Time spent meeting with suppliers = 2 hours
Balance time to do other work (preparing report(s), scheduling, inventory, monitoring plant and oil quality, phone calls, communicating with downlines etc.) = 4 hours

To tell you the truth, 4 hours are sincerely not enough to do or complete the jobs. That's why most of the time I have to go back late. But I don't want to go back very late because the rate of work productivity decreases as the day turns dark. And I live 50 km away.

"Science can amuse and fascinate us all,
but it is engineering that changes the world."

Isaac Asimov
(American writer, professor of biochemistry)

Plant 5S

I have my own plans and targets for my plant. I want to make the plant a better place to work. I want to improve the system and way of work. I want to beautify the 9 storey physical refining plant. I want all the pumps, heat exchangers, insulation, vessels, cable tray, etc to look nice and well maintained. I want the paint on walls to be fresh and free from dirt. I want the plant to be totally clean and shining. I want the forklift to be in excellent tip top condition with new paint. I want to see the cooling towers, bleaching earth store and spent earth area to be in superb condition. I'm working on all of that with the help of all the plant supervisors and plant operators. These are what we call 5S housekeeping. 5S is a Japanese Concept of Total Productive Maintenance. I hope to achieve them within the second quarter this year.

5S is a reference to five Japanese words that describe standardized clean-up:

Seiri: *tidiness, organization*. Refers to the practice of sorting through all the tools, materials, etc., in the work area and keeping only essential items. Everything else is stored or discarded. This leads to fewer hazards and less clutter to interfere with productive work.

Seiton: *orderliness*. Focuses on the need for an orderly workplace. Tools, equipment, and materials must be systematically arranged for the easiest and most efficient access. There must be a place for everything, and everything must be in its place.

Seiso: *cleanliness*. Indicates the need to keep the workplace clean as well as neat. Cleaning in Japanese companies is a daily activity. At the end of each shift, the work area is cleaned up and everything is restored to its place.

Seiketsu: *standards*. Allows for control and consistency. Basic housekeeping standards apply everywhere in the facility. Everyone knows exactly what his or her responsibilities are. Housekeeping duties are part of regular work routines.

Shitsuke: *sustaining discipline*. Refers to maintaining standards and keeping the facility in safe and efficient order day after day, year after year.

Plant Shutdown

The followings are my journey/diary during a plant shutdown in January 2007.

Day #1

My plant has to stop for shutdown maintenance work today, 4 weeks earlier than the expected date. We have to entertain the sudden shutdown because lack of CPO. I was instructed by my boss to come to work in afternoon and stay until night. This is to supervise and monitor the plant shutdown, cleaning, maintenance and plant startup activities. My senior colleague will cover the normal shift. I'll be working like this for at least one week. Its ok, I don't mind about it. So far the plant stoppage run smooth and is few hours ahead of schedule. Some of the manholes of the vessel have been opened just now. From the steam test, we found 10 leaking points that need to be rectified, either by welding or replacing the portion/pipe that cannot be welded. I've been recording all the activities and will produce a complete shutdown report for the management after the plant shutdown ends.

Day #2

Today is the second day the plant is on shutdown. There are really lots of work to be done, but less number of manpower due to the public holiday. Yesterday we are already in front of schedule. Today, we are already behind schedule. I'm quite disappointed with the whole progress. I'm upset with the work organization of certain group involved in the shutdown work. However, I hope tomorrow, we can speed up everything and catch up with all the pending works.

Looking at the plant shutdown from a different perspective, personally, I learned lots of things. Every shutdown is unique. There will be a lot of common jobs, and some changes, or some new things/project/modification to be done. I really cherish and appreciate those experiences, and it definitely enriched my technical knowledge and experience in managing the plant shutdown.

Today, among important jobs that were carried out:

-Steam test - at the beginning of shutdown to detect leakages.
-Hydro test for High Pressure Boiler - to detect leakages.
-Opening manholes & dismantling flanges - common jobs, in fact, I assisted
to open one of the manholes. Super tired (20 bolts + nuts size 24).
-Shell Tube cleaning - via Cleaning in Place (CIP).

Day #3 & #Day4

Today, I would describe the plant shutdown job as less head ache. Everything
seems to run smooth. Thank God. Everybody knows what to do.

Only one thing that shocked me in the early hours (6am). My operator called
& told me that pressure for HP boiler hydro test has reached 52 bars. The
water feed pump has been stopped. However, the pressure keeps on
increasing gradually. I already instructed them to control the pressure between
50 – 60 bars. The increasing pressure worried me. I don't want the pressure
to reach above 70 bars without enough manpower observing the situation.
After thinking, I called and instructed him to monitor the pressure closely. If
the pressure exceeds 67 bars, slightly open/throttle the venting so that the
pressure would not increase. I choose 67 bars simply because the HP boiler
always runs between 64-66 bars. Therefore, it will still be safe to operate or
control the pressure at 67 bars.

The plant operators informed me that the increasing pressure might be the
effect of conducting the hydro test and the shell and tube CIP at the same
time. The CIP temperature is maintained at 80ºC. The heat from the CIP has
exerted some pressure to the other activity, which is the hydro test. This is
because, in normal plant operation, steam from the HP boiler will transfer
heat to the oil in the shell tube heat exchanger. Therefore, we cannot do the
hydro test. We have to give way for the CIP first. Then only we can proceed
with hydro test to avoid the pressure build up as earlier.

Day #5

Time is running swiftly. We stopped the plant on Monday, and now Friday is over. In day #5, we had the JKKP (Department of Occupational Safety & Health - DOSH) inspector inspecting all our pressure vessels and high pressure boiler. Everything was smooth and working perfectly fine. Immediately after the JKKP inspector left, we boxed-up the vessels and high pressure boiler. We fixed back the Niagara filter leaf. We did some cleaning as well. Tomorrow, we are going to complete remaining maintenance job. Later, we will carry on with steam and air test. Finally, we shall start up the plant. Hopefully we'll have a perfect start up.

"Focus your efforts 100% in stuffs within your control,
focus 0% in stuffs beyond your control"

Zaki Yamani Zakaria
Founder of Chemical Engineering World Blog &
Chemical Engineering World FB Page

Plant Shutdown – No Water Supply

Plant Shutdown activity is going on which means we need lots of water for cleaning. In addition to that we have to conduct CIP for the shell and tube heat exchanger. CIP is carried out to remove the scale and carbon deposited inside the shell and tube in order to improve the heat transfer. Unfortunately, suddenly the water supply was cut off without any notice. Worse than that, nobody realized the unavailability of the water supply for about 3 hours. Upon realizing it, we have to depend on our reserve water supply which is only about 1500 m^3. The reserve water is normally saved for the utility boiler because steam is a necessity to run a plant. If steam is not produced with sufficient pressure, production will be affected and if too less, we will be forced to stop other plants from operating.

A decision was later made by the manager. All activities involving water had to be stopped. We have to save the water for the boiler. The plant shutdown cleaning activity was affected and delayed as well. Luckily, the water supplied was back a few hours later and we continued with our shutdown cleaning activities.

"The way to succeed is to double your failure rate."

Thomas J. Watson
(Pioneer in the development computing equipment for IBM)

Air Test before Plant Start-Up

The excruciating plant shutdown was finally over on Saturday. We started-up the plant on Saturday 2200 hrs. Hopefully everything will run silky smooth. We conducted both air and steam tests twice in two different sections in the plant. From the first air test (at the bleaching section), we hold the air pressure inside the vessels and pipelines at 1.5 barg (it took about 1 hour ++ to build the pressure up to 1.5 barg)*. The pressure dropped to 1.0 bar in 1.5 hours. This indicates that there were some leakages in the system and we need to search for it. Supervisors and plant operators searched for the leaking points, equipped with alkaline solution to detect leaking air by spraying them on the vessels and pipelines. It's like finding a needle in a haystack, the leak could be anywhere on the pipe surface. The plant is massively huge. After numerous sprays and long search for it, we detected quite a number of tiny bubbles. Leak can simply be detected when bubble appeared from the leaking points. They were marked and maintenance fitters attended and welded them after we released the air.

Releasing air process does not take a short time. That's why we need to coordinate the shutdown properly and effectively. After detecting the leaking points, air was released from the vessels and pipelines. This took about 2 hours. After air pressure is lesser, maintenance fitters welded the leaking points. Upon completion and satisfied with the workmanship of the maintenance team, we repeated the air test, hold it at 1.2 bar. We targeted the previous leaking points, sprayed them and they were all ok except for one point beneath a retention vessel. We released the air again at that particular section (deodorizing section), called the fitters and asked them to re-weld the point. There goes another hour, waiting for the air to escape. At that point, bleaching section is ready for plant start-up.

*The target pressure for air test may differ from plant to plant and also depends on executive decisions.

Plant Shutdown Steam Test

I came home from work and it was already midnight. I like it this way because I can escape the heavy traffic jam during the peak hour. It was quite a hectic day at work especially when the shutdown just started. We have to do a lot of activities to stop the plant. I experienced and learned a lot of things.

We did the steam test to check and detect leakages in the vacuum pipe. This is because the vacuum system in the plant is poor. We build up the pressure to 1 barg and inspected the line. It's easy to view some leakages because it is simply very obvious since steam came out vigorously from it. While we already settled with the number of leakages that we found earlier, we discovered a few more very tiny leakages that is impossible to see. They were extremely small and appeared like a crack. Very small droplet of water appeared out of the pipe after the steam test was set up to 1.8 bars**. We marked the leaking points with red colour spray. The maintenance fitters will later weld or clamp or replace the leaking point tomorrow.

I hope after the entire discovered leaking pipe problems have been solved, the vacuum system will be improved and the plant will be in a better shape than before, as good as new. I hope so.

Steam test is almost similar to air test. Instead of injecting air, steam is injected into the system. The reasons for performing steam test is for swifter leakage detection (around the pipes, insulated pipes and vessels) as it can be visually observed (steam will visibly shoot out from the leaking point) and produce a loud screechy noise. Application of steam test also works perfectly at high altitude (refinery plant can be up to several meters high). Although steam test have its own advantage, applying more steam will only be wasting time, energy and money. In addition to that, it will a take longer time to release the steam after completing the process.

**The target pressure for steam test may differ from plant to plant and also depends on executive decisions.

For our case, we did steam test at the beginning and at the end of the plant shutdown. Both were necessary (depending on the condition of the plant). We identified the leaking points and welded them from the earlier test. The final steam test was carried out to double check before plant start-up. Both steam tests have their own downside. The first one will delay cooling of vessels, because steam is hot. Therefore, upon opening the manhole, we have to allow at least 1/2 a day before entering the vessels. Another downside which applied for both earlier and later steam tests were the amount of water produced as condensate from the steam. During plant cleaning, existence of water is not really a problem. However, during plant start-up, water has to be fully drained before pumping oil into plant. Having a mixture of water and oil will create various problems including oil quality.

"I was originally supposed to become an engineer but the thought of having to expend my creative energy on things that make practical everyday life even more refined, with a loathsome capital gain as the goal, was unbearable to me."

Albert Einstein
(Influential theoretical physicist)

Accident and Injury

It is still day #3 and I was walking towards my plant after passing by the Low Voltage (LV) room. It was a rainy wet night. Suddenly, in split second, I slipped and fell down. My left foot hit a 1 cm thick metal plate before sliding into a drain beside it. I sat there for a while helplessly. It really hurts. I watched around to call for help but nobody was at sight. I inspected my left leg and to my surprise, I have 2 really bad cut/wound and they were big! Blood slowly appeared from the deep wounds. I rose and walked slowly to the washing basin nearby the control room and washed the wounds with a lot of water, washing the blood a way. I ran the water passing through my leg for about 5 minutes. By that time, my subordinates had surrounded me and checked out my leg. It really looked bad. It definitely had to be stitched. Both of my feet were shivering. My whole body was also shivering. KC, my colleague, drove me to the nearby private hospital. The medical assistance inspected the severity of the wounds. He injected me with antibiotics before sending me for X-ray. It almost hit my tibia bone. They want to check if the bone was fractured or not. Luckily it was ok. Then I have to overcome my worst nightmare, to be stitched. Altogether, I have 10 external stitches and 6 internal stitches. This is indeed one permanent memory.

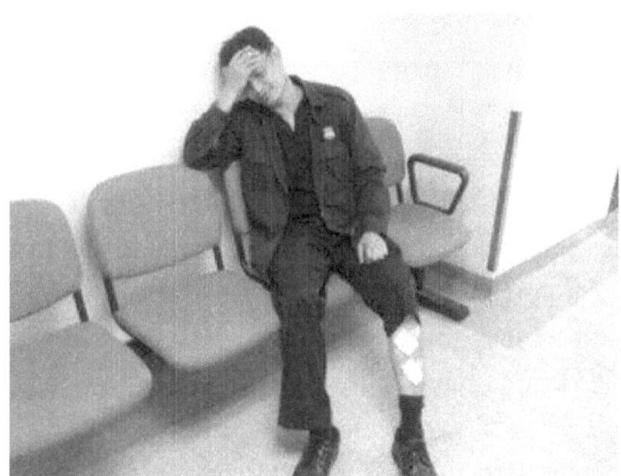

I was at the hospital, waiting to be treated. Later I received several internal and external stitches. The person that was treated next to my bed in the emergency room was pronounced dead when I was there and I witnessed his family shouting and crying. Oh what a terrifying moment and experience.

Another Plant Shutdown

Today, I was informed that we have to stop another plant that I'm in charge of for maintenance 3 months earlier than the exact schedule. The reason was simply because it's going to be Chinese New Year festive season which means less crude palm oil (CPO) will be received due to tankers not allowed to be on the road. At the same time, the brutal rainy season since last December created flood in most territories in Johor which gave direct impact to the supply of CPO. A lot of palm oil plantations were flooded and at the same time people cannot harvest the palm fruit.

So there goes another public holiday! Few weeks ago I could not enjoy the New Year and Hari Raya Haji because of a plant shutdown too. Well, this is something to be expected if we work in this kind of industry.

Cooling Tower Problem Again!

Just now, I was informed by my supervisor that the clean cooling tower water has turned into a milky coffee colour! Oh No....this could not be happening again....I went to the cooling tower and check out the water. It is true. The water colour has changed from clean colourless water to a milky coffee colour. This is going to be a very bad nightmare for me. The last time this cooling tower had a problem (oil leak into it), it took about 6 months for it to recover and we had to carry out a major service on it. I hope this time; it is not some unwanted oil leaking from the heat exchangers from 4 different plants. Well, at least no slime is observed yet. Or, perhaps, the slime has not yet grown. I need to inspect it again tomorrow morning. As for today, the best thing we can do was removing some scum floating on top of the cooling water surface, did some blow down and some sand filter backwash. Hopefully it can dilute anything that contaminates my cooling tower water.

Cooling Tower Water Quality Disaster

Today, the cooling tower water quality does not change much. It is still about the same colour as yesterday. Scum and foam began accumulating on the surface of the water. It is not a nice scenery. It looks terrible. Some foam and water sample was taken yesterday and sent to the lab for analysis. The lab tested the IV (iodine value) of the oil extracted from the foam and water. The IV was 100+. That gave a very clear indication of what oil is leaking into the cooling water system and from which heat exchanger and plant too. It was soya bean oil which has that very high IV. Palm oil and coconut oil which was running in my plant have IV of about 52 and 9, respectively. Another plant was running soya bean for about 3 days already and the plate heat exchanger in that plant is suspected to have internal leak. Tomorrow, fitters from maintenance department will dismantle and replace the gasket for that problematic heat exchanger. Let's see what's going to happen tomorrow.

Cooling Tower Improvement

Luckily today the cooling tower water condition is improved. I can see the water quality getting better and clearer. Thanks to some aggressive blow down and extra backwash via the sand filter. The early detection of which heat exchanger contaminated the cooling system helped a lot in hindering the leaking of soya bean oil. That heat exchanger has already been straight away quarantined and dismantled. Just now, the plates were immersed in hot caustic to clean and remove scale sticking on it. Tomorrow, gasket will be clipped onto the plates and the heat exchanger will be re-installed. Hopefully, the heat exchanger will be ok and would not leak internally or externally. I also hope by Saturday, the cooling tower water quality will be perfectly fine and ready for operation.

Hydrocarbon Leak Detection in Cooling Water Systems

It is our hope to have a smooth cooling tower operation. However, once in a while we will be surprised with problems such as leaking oil in the closed system. Hydrocarbon leaks in refineries and chemical plants can quickly cause fouling and outbreaks of microbial growth. This, in turn, results in rapid loss of heat transfer and system efficiency. Detecting, locating, and stopping these leaks as soon as possible is critical.

A hydrocarbon leak detection procedure is used to pinpoint process equipment that has oil leaking into a cooling water system. In the past, this testing has been a laboratory procedure and could require considerable time to obtain results. Newer test procedures can be performed in the field and provide immediate on-site results.

Samples of water from the inlet and outlet of a suspected leaking unit were collected. When an increase in hydrocarbon level is found in the outlet sample, the leaking unit has been identified. This is the culprit. It needs to be isolated. It definitely needs to be treated. The longer the leak continues, the greater the possibility of equipment damage or plant shutdown.

Technology is a gift of God. After the gift of life it is perhaps the greatest of God's gifts. It is the mother of civilizations, of arts and of sciences.

Freeman Dyson
(English-born American theoretical physicist and mathematician)

Cooling Tower Heart

The heart of the cooling tower is the "Film Fill Media" or "Infill". It will experience thermal shock, atmospheric dust as well as slime and algae growth. These are the phenomenon that we can expect when taking care of a cooling tower. The infill will become brittle and be disintegrated (wear and tear) along its shelf life. The potential contaminated particles will cause the cooling system efficiency to drop due to the cooling tower clogging. When this happens, a long repair / down time of the cooling system can lead to very high losses on operation cost as well as production cost.

With a proper Preventive Maintenance Program for the cooling tower which includes replacement of infill, the cooling efficiency will be dramatically improved, saving a huge amount of electricity energy consumption, longest and most dependable useful system for cooling system. There are so many things to be learned while handling a cooling tower.

"The human foot is a masterpiece of engineering and a work of art"

Leonardo da Vinci
(Italian polymath of the Renaissance)

Heat Exchanger Plate Condition

After we use a plate heat exchanger for quite some time, let say a year for instance, you can expect some fouling on the heat exchanger plate oil side. The steam side might not be as dirty as the oil side. When fouling is formed on the surface of the heat exchanger, the heat transfer is no longer effective. If you calculate the overall heat transfer coefficient, the value will be lower compared to when you first commission the same heat exchanger. Therefore some cleaning is necessary. Just choose either by dismantling the heat exchanger or perform CIP using de-carbonizer or hot caustic.

This is a clear example of bad fouling taking place on a plate heat exchanger oil side. Have you seen something like this before?

Overall Heat Transfer Coefficient (U-Value)

I've been collecting data for the heat exchangers in my plant. There are 7 plate heat exchangers (PHE) that I closely monitor. There are several more that I do not really focus in. Among the data collected are: inlet & outlet temperatures, flow rate and surface area. My aim is to calculate the "Overall Heat Transfer Coefficient (U-Value)" so that I can know which PHE has become heat transfer inefficient. However, after a series of data, I observed that sometimes the PHE have high U-value and sometimes the same PHE have lower U-value. And I make a conclusion for myself that the U-value in my case is actually strongly influenced by the flow rate. In my plant, sometimes we have to increase or lower the flow rate due to some quality control issue. Of course, for me the effect of Log Mean Temperature Difference (LMTD) is not as great as the flow rate effect.

"The meeting of two personalities is like the contact of two chemical substances:
if there is any reaction, both are transformed"

Carl Jung
(Swiss psychiatrist and psychoanalyst who founded analytical psychology)

Butterfly valve

In a production plant, we need a very good process control equipment and system. We learned the theory in Chemical Engineering subject called Process Control and Instrumentation. It was hard for me to imagine what process control and instrumentation is all about when I studied the subjects few years back.

When I started working in a plant, then everything becomes clear. One of the most important equipment is valve. There are various types of valve. We have butterfly valve, ball valve, gate valve, globe valve, check valve etc. Each type of valve has its own pros and cons as well as functional area. There are a lot of things to talk about valve. However, in this post, I'm going to cover just a little bit about butterfly valve.

The following photos illustrate a typical butterfly valve. It is a 3" Belgium Ventiel (BV) butterfly valve. This valve had created a very serious contamination and loss earlier because it leaks.

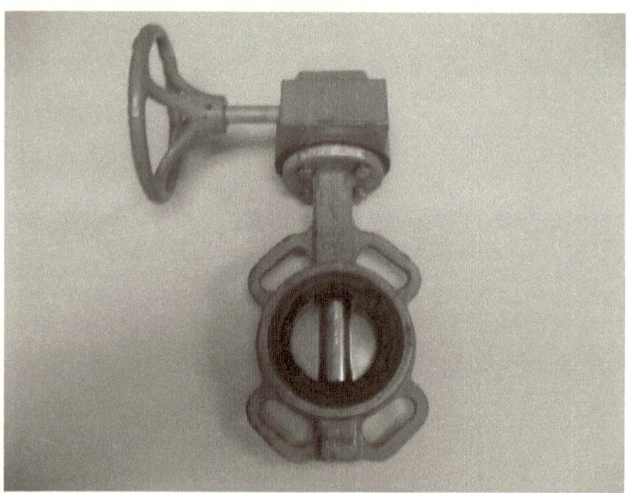

Typical butterfly valve - gear type. The gear is used to open or close the disc. The disc is made of stainless steel. It is now on close position.

Closer look of the butterfly valve when it is fully open, allowing liquid to flow passing through it.

The arrow shows that the seat / liner is damaged and worn out. This is one of the reasons leakage can occur from a valve. There are various types of seat / liner to choose for different types of liquid and process conditions. The disc is in half open position.

Different angle of the butterfly valve.

"*I have not failed, but found 1000 ways to not make a light bulb*"

Thomas Edison
(American inventor and businessman)

Wet Wet World

It was day 4 of the plant shutdown and it's nearly coming to an end. It was also a wet wet world in the plant. We have to do a lot of washing and cleaning. Tomorrow will be the Department of Occupational Safety & Health (DOSH) inspection day and we have to ensure that the plant and pressure vessels are perfect and clean. Just now, we removed scrap metal, spent earth, damaged insulation and other waste / rubbish. Darkish and dirty floor with oil traces was mopped and cleaned. In general, I was pleased with the work performed by my down line.

"I consider nature a vast chemical laboratory in which all kinds
of composition and decompositions are formed"

Antoine Lavoisier
(French nobleman and chemist who was central to the 18th-century
chemical revolution and who had a large influence on both
the history of chemistry and the history of biology)

Electrical Problems and Preventive Measures

To run a plant, electricity is one of the most important utilities besides water and natural gas. Electricity is required to run the motors, pumps, control panel, control system, lightings, cooling tower fans, etc. If electricity is suddenly cut-off due to whatever reasons, the plant will immediately be affected. Pump would not run and all valves must be immediately close to contain the oil. Vacuum system will fail and oil quality will be off-spec, hence the oil must be rejected to the crude storage tank. All factory or plants spend tremendous amount of electricity to run a plant besides using it for basic lighting, air conditioner, PC, server etc.

I am not very good in electrical subject but I have to know and learn few of the basic electrical stuff because it is required to run a plant. I learned about it from the charge man, electrical executives and supervisors. Here I list down some of the major electrical problems that normally would occur. At least it can give us some idea on what potential problem can occur from the electrical perspective.

Major Electrical Problems:
(A) Over-Current
(B) Earth Leakage
(C) System failure due to lack of maintenance

(A) Over-current
Cause of over current:
(1) Excessive heat generated: additional equipment installed without upgrading panel board and cable.
(2) Motors / Equipment Failure
Motors that are not maintained properly require more power/current e.g. air conditioning system that is not serviced may demand more current.

Maintenance Wise

1. Maintain the equipment to make sure the current demand is at rated capacity.

2. Regular maintenance and cleaning - Visual checking and Infra-red temperature checking (annually)

Over-current protection & Prevention Equipment

(1) MCB (miniature circuit breaker)

(2) MCCB (moulded case circuit breaker) – for current > 100 A (e.g. motors)

(3) ACB (Air circuit breaker) - > 400 A (main board)

(B) Earth Leakage
(1) All electrical equipment must be properly earthed

- To protect people from being electrocuted

(2) All leakages must be brought down to a safe limit

(3) 3 types of earthing relay systems:

(i) EF (earth fault) relay (> 200A, high tension & LV system)

(ii) ELR (earth leakage) relay

(iii) Earth leakage circuit breaker (<60a>

(C) System failure due to lack of maintenance

Precautions:

(1) Tighten loose connections
(2) Regular calibrations of equipment
(3) Regular cleaning of equipment

Paint in Plant

I used to think painting was simple and easy. But, when it comes to painting pumps, heat exchanger frame, walls, beam etc it opens a new chapter of knowledge inside me. Yesterday, we met with Nippon paint supplier. We learned something new about metal surface, epoxy, coating technology and others.

As I mentioned earlier in my previous post regarding 5S, we intend to improve the cleanliness and condition of our plant. We want the paint to last for about 2-3 years minimum. We don't want the paint to be nice for only a few months' time before it faded and chipped of.

Initially, the supplier said that zinc chromate should be applied to prevent corrosion from progressing on the metal surface. It acts as a primer (which means the first layer to be applied on the metal surface). Then, undercoat or finishing coat will be applied and this depends on how much cost we are willing to pay (either to have single coat or double coat). Having 2 layers of coating is better but the price is higher.

However, after considering the high temperature of pumps and heat exchanger frames and other tough environment factors in the plant, a primer layer of epoxy was suggested. Earlier, I heard that epoxy as a primer is the best to suit for tough environment, but technically I don't really know the chemistry detail behind it. A special coat (that will create stronger bond with the epoxy primer) will ensure that the quality of paint and its colour will last longer. That combination will be slightly more expensive due to its quality.

We did get some advice on the correct painting technique and other relevant information. Now, I'm waiting for the quotation of the primer and coating (paints).

Following is the correct painting technique from Nippon.

<u>Painting the Right Way</u>

Now that you have chosen your ideal colours and type of finish, here are few important tips to make your paint job came out with flying colours.

1. Prepare the Surface
   This is the most important stage. Ensure all surfaces are clean, dry and grease-free. Wash with water & liquid detergent. For heavy and stubborn stains, use turpentine and wash thoroughly. Fungus & mould: remove as above, followed by fungicidal wash or bleach.

2. Apply the Sealer or Primer

3. Apply the Undercoat

4. Apply the Finishing Coat

5. Be responsible to the environment

Do not dispose of unwanted paint down the drains.
Allow paint to dry out in the container before dropping them into the rubbish bins.

Be Cautious at Dirty Cooling Tower

If you're a in the oils and fats industry and are working in a refinery, there are clean cooling tower and dirty cooling tower. Clean cooling tower, by its name itself, is clean while dirty cooling tower, from its name is not that clean. Clean cooling tower is mainly to cool down the refined bleached deodorized palm oil (RBDPO) in the heat exchanger and for several other purposes; and it is a closed system. The dirty cooling tower, on the other hand is much required in the vacuum system, rather than an open system and always end up with palm fatty acid distillate (PFAD) inside it, before temporarily staying in a hot well. As a result the water become dirtier and when it arrives at dirty cooling tower, all this unwanted things accumulates.

The condition of this dirty cooling tower is not very pleasant. The water is already mixed with PFAD and some trace of oil. You need to be extra cautious when dealing with this type of cooling tower. One thing that you must avoid the most is welding. DO NOT WELD anything nearby the dirty cooling tower. This is very dangerous and hazardous. The fatty acid and oil that is contained in the cooling tower water is very flammable and can catch fire easily. So, warn your people/technician/fitter > Never deal with any type of fire/spark nearby the dirty cooling tower.

"Habits are qualities of the soul"

Ibn Khaldun
(Fourteenth-century forerunner of the modern disciplines of
historiography, sociology, economics, and demography)

Steam Trap Failure Issue

As I read through the "Expert Q and A - Timely Detection of Steam Trap Failure" post in the Chemical Engineering Blog, I was called to share my opinion and experience. This is because I also had similar problem in my plant regarding maintaining the steam trap functionality. (p/s: For those who don't have an idea what is steam trap and how does it work, and what is the importance of steam trap, google and read about it). In the post, a question was asked:

Q: A steam trap tested as operational during an annual survey can fail at any point after the test. These failures go undetected until there is some type of system failure or until the next time a steam trap survey is performed. What is the most effective approach to detect failure and maintain a best-in-class steam system?

Tracy Clupper from Armstrong International (Specialist in utility system solution) responded and provided an answer which from my opinion is good.

A: When evaluating methods of testing steam traps, it is important for companies to be realistic about their current needs and capabilities. A company with plans to implement an annual in-house trap-management program with no dedicated or trained personnel is wasting time and money. Hiring a third party to test traps and provide a condition and savings report can be a more effective way to manage a trap population when in-house labour is not available. This approach, however, does not solve the problem of system-related emergencies due to undetected trap failure, nor does it afford the ability to truly capitalize on energy-related savings. One failed-open, high-pressure drip trap can cost a company thousands of dollars in annual steam losses. These losses are shown on a static report as a function of a trap that has failed for an unknown period of time. There is no cost avoidance in this situation, only the potential to stop losses that have already occurred. In addition, the impact of high-pressure steam traps discharging into lower pressure parts of the condensate system can be detrimental to the overall efficiency of the entire system.

A 24/7 monitoring system can more effectively detect and identify points of failure for immediate maintenance response. Both hard wired and wireless systems are available on the market. The key to implementing any continuous monitoring system successfully is to identify the costly, critical or dangerous areas of the steam system and target those areas first. Then evaluate the continuous monitoring system's reporting capabilities; keeping in mind that immediate notification of trap failure is critical, but savings validation is also important toward proving the monitoring system's effectiveness toward true cost avoidance.

Here I just want to add up some points to what Tracy Clupper has mentioned above (which I added in the post comment section).

Initially, in my plant, there was nobody monitoring the steam traps. As a result, the steam consumption increased and our vacuum system became inconsistent. After realizing the situation (effect of steam trap failure), we assigned one manpower (from maintenance department) to inspect and record the steam trap temperature inlet and outlet every 2 months. We can know the steam trap is failing if the inlet and outlet temperature is different. However, this is still not enough, because the steam trap may fail in between those two months. Therefore, I asked my supervisors to check all the steam traps every time they do the round in the plant during their shift. They don't have to carry with them a temperature gun to check the temperature. They just need to feel the steam trap inlet and outlet by their bare hand (just a touch and it's not dangerous!). If the outlet is colder than usual, that means the steam trap is not working!!! This is because steam is not flowing through. It is a simple practice; however, ensuring them to continuously do the inspection on a daily basis is another challenge. I'm thinking of making a checklist, so they can follow the checklist and feel all steam traps available in the plants, and have the black and white record.

Hiring a third party is also a good option, but why need to pay them if we can enforce or ask our own people/staff to do it. Those are just my opinion about the steam trap failure issue.

Wooden Cooling Tower Drift Eliminators Got Dirty

The following photos illustrates the cooling tower infills and drift eliminators. That's why after some period of time we need to clean the infills. In worse cases, we need to replace the infills. In this example, the infills are made from cengal wood.

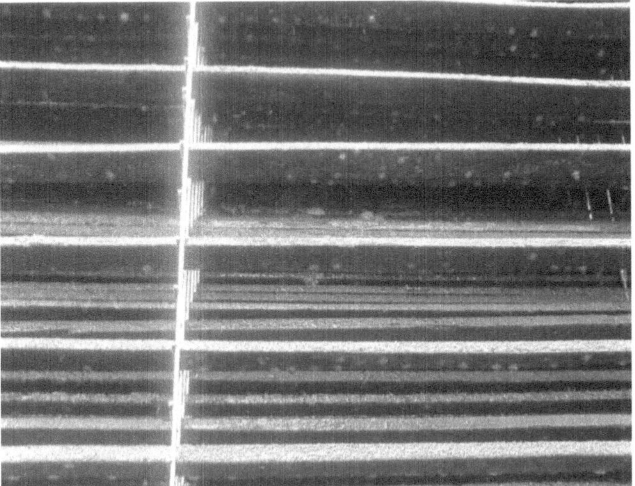

Wet cooling tower infills

Dirty and crooked drift eliminators. These drift eliminators were then replaced with a new set.

The infills arrangements are important to create droplets of hot water dropping from the top. The water droplet (which is hot) will then progressively be cooled down by the air which is sucked passing through the infills from outside by the cooling tower fan.

In actual fact, there are many components in a cooling tower. Each and every component needs to be in good working condition to ensure smoothness of the whole operation. If a cooling tower is dirty, it could not effectively perform its duty to cool down the water temperature, normally to decrease 10°C below the designed incoming water temperature from plant.

I am glad that I have worked closely with several cooling towers performing various tasks such as taking care of the water treatment program, cleaning, replacing drift eliminators and infills, servicing target nozzles and water distributor above the cooling tower deck, checking, servicing and maintaining the cooling tower fan and pump, and many more. I just realized after listing all these that I have done many things around the cooling tower ☺.

ISO and HACCP Audit

In November 2007, my plant was internally audited. The auditors checked our ISO documentation (ISO stands for The International Organization for Standardization) and Hazard Analysis Critical Control Point (HACCP) documents thoroughly. Few of the GMP (Good Manufacturing Practice) documents were reviewed as well. It took the auditors 4 hours to go through some of the documents. Besides interviewing me, the auditors interviewed the plant supervisors as well. By having an internal audit like this, we are continuously improving our documentation system. At the closing meeting, we were awarded 3 Non-conformance (NC), for not complying with the ISO and HACCP standard. I aspired to take appropriate corrective action towards the NC's charged to us.

The next day, I was an internal auditor to audit two departments in my work place. One must be a trained auditor to carry out an audit. For that, today, I have made early preparation by reading that department's ISO document. It's quite a new field to me, but at least I'm learning something non-technical from other department's operation and documentation.

An important thing to take note is, whenever, we charge a Non-conformance (NC), we must include the proof and associate the non-compliance with the right clause from the ISO or HACCP standard.

Becoming an Auditor

I was instructed to audit other departments in my work place. I was actually being trained to become an internal auditor. Yesterday, I audited two departments: "Consumer Packing Marketing Department" (CPM) and "Shipping Department". I have read both departments' procedure manuals and quality objectives earlier. CPM documents are easier to comprehend and digest compared to the confusing terms and definitions for the shipping department. Then I wondered, why did they send a process engineer like me to audit this shipping department? I knew it was not going to be easy. I don't even have any shipping background. Well, instruction was given and I have to perform the job.

Yesterday, to make things worse, the lead auditor, who happened to be a senior employee, could not join the auditing process, leaving just me and another colleague from account department. In the morning, the two of us headed to the CPM department, began the audit opening meeting and on-site document checking. It took us about 3 hours flipping and checking documents besides asking questions. The audit process went smooth, the auditees were very cooperative and easy going. The CPM department job scope and flow chart is straight forward and easy to understand. After completing the audit process, we prepared the report and conducted the closing meeting. No non-conformance was given. We just gave them 3 observations that they need to respond and improve with respect to some minor poor documentation control.

Later in the afternoon, we proceeded to the shipping department and met the auditees (3 people whom happen to be senior executives - 2 versus 3!!!). Deep inside my heart, I already knew that this is not going to be easy simply because I have almost no knowledge on shipping stuff! I just read the quality objectives and manuals but hardly understood them. We began the audit process and I requested for their quality objectives (they have 3 objectives). I asked for the quality objectives but got confused with all the terms, definitions, phrases, concepts etc. There were few quality objectives that they could not achieve, but they have no control of the problems. It was due to some external factors, which again, I barely understood. Then I flipped through some of their documents and checked them. At least, I noticed something that is worth an observation. At the closing meeting, we informed the shipping executives about the only observation discovered and asked them to rectify it.

Oh boy, I was glad that the audit process was over. I felt like a fool auditing the shipping department. However, I took it positively. I knew it was a priceless experience and I managed to learn various activities in CPM and shipping department. At least now, I have a better idea on how those departments operate. The experience enriched me to become a better auditor, besides my main job - process engineer.

"Pray to God, work smart and hard, tawakal.
Repeat them until you taste success"

Zaki Yamani Zakaria
(Founder of Chemical Engineering World Blog &
Chemical Engineering World FB Page)

Fixing Plate Heat Exchanger Gasket

The following photos are Alfa Laval heat exchanger plates during gasket installation. Learn more about Alfa Laval plate heat exchanger.

Plates are arranged piece by piece after fixing the clip on gasket.

The black clip on gaskets can be seen beside the plates. Clip on gaskets do not need glue to be positioned on the plates.

Overview of the plates. The plates are clean and looks like a new set of plate.

These are the clip on gaskets before they're fixed on the plate.

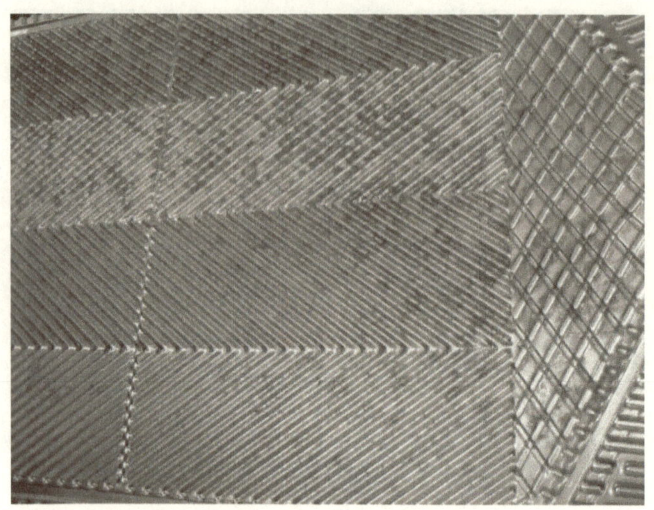

This plate is not really clean. Can you see the brownish colour on the plate. That is the stain after cleaning the plate that won't easily come out.

"Water is the driving force of all nature"

Leonardo da Vinci
(Italian polymath of the Renaissance)

Being Well Informed

As an engineer, one of the most important skills necessary is the art of being well informed. We must always be aware of everything that is going on inside our zone and our area of responsibility. We don't want to be the last to know about any occurrence. We don't want to be alerted about any problem by our superior. If that happens, it simply means that we are not carrying our job as an excellent engineer and care the least about our responsibilities.

Sometimes it makes us look better if we pretend to know about the problem in front of our boss when asked. However, the cover up must be good and we better get the answer right. I remember when I started working few years back, my boss asked me about the status of our shell and tube heat exchanger cleaning in progress (CIP), and I answered him. That time, I pretended to know what he was asking and answered him, hoping that my answer was OK. Unfortunately, he scolded me and said to my face that I don't know what I was talking about. Then he went away, obviously pissed off with me. Since then, I tried to know and learn as fast as possible. I must always be ready and able to answer any questions thrown to me by my boss.

"At its heart, engineering is about using science to find creative, practical solutions. It is a noble profession"

Queen Elizabeth II
Queen of the United Kingdom

Sketch Drawing

Understanding a PID diagram might be tough for beginner. However, after knowing the symbols and other basic drawing information, you can get hold of it. I've seen a lot of drawing. I'm also upgrading and improving my plant's drawing (in process).

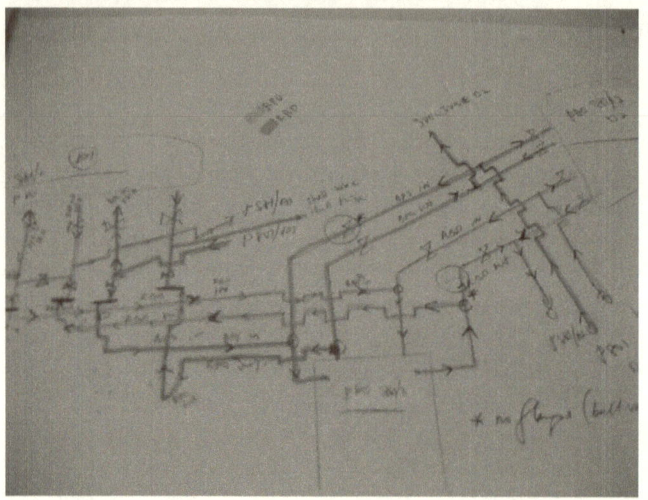

The photo I include above is a drawing/sketch of pipings connected to few heat exchangers from two plants. It's just a simple fast drawing prepared by my supervisor for us to understand the piping layout connected to some plate heat exchangers. Looking at the original diagram, it's quite difficult to comprehend. I have to apply some colours to have better visualization of the layout.

Leaking Pipeline

There are few things we need to know about our pipeline and our plant process. Pipelines are very important to transfer fluid (eg. oil and water), gas (natural gas and steam), or solid (bleaching earth and other powder form solid). Selecting the right pipeline material, right sizing, and right schedule is imperative in ensuring smooth and perfect production operation. Sometimes, we have to be very careful, because, although we have selected the correct material, size and schedule, unfortunately the supplier cheated and supplied defective pipeline unrealized by us. As a result, the pipeline does not last long as expected. Crack appears and progressively developed into a bigger crack and hole. If oil is moving with some pressure in the pipeline, we can immediately detect the leaking oil when it dropped.

It is very clear and if the leak is big, it will be very messy and dirty situation. What if the pipeline is a vacuum line? We can't see anything. We can only detect pressure vacuum dropped in the control system via the pressure transmitter installed in the line. It's impossible to weld the hole because the pipeline is running under vacuum. Any spark will be sucked into the system and will create fire and explosion.

Most of the time, we cannot weld the pipeline that is in operation. The condition shown in the photos are situations that were done for temporary solution and we need to wait for the plant to shutdown to weld. Hence, before any plant shutdown takes place, we will have a very extensive list on plant jobs to perform.

This 12" pipeline has some very small leak that affected the vacuum pressure. The leaking points were covered by another piece of metal to block the holes. However, it still leaks (indicated by the yellow sign).

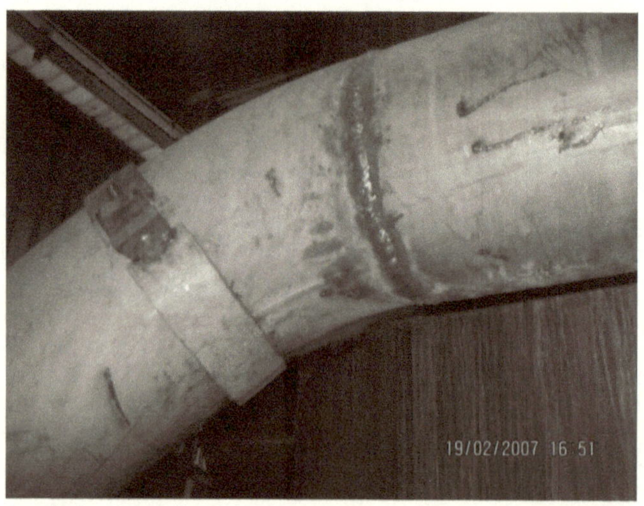

Similarly, this 12" SS316 line also has some tiny pin hole which is very hard to see by naked eye.

Partially Blocked Pipeline

This is one of the reasons a plant shutdown is important. This pipeline is partially blocked with spent earth mixed with gum. During plant shutdown and maintenance, we can clear and clean the line.

Can you see how terrible the condition is inside this pipeline?

Do I Enjoy My Job?

I received an interesting question (which is this post's topic) from anonymous in the comment area in my previous post. I shall answer them based on my current job (oil and fats industry). In future post, I shall reveal what I feel being in the oil and gas industry.

In the oil and fats industry, now, I take care (with a senior colleague) of a big physical refining plant processing the likes of palm oil, coconut oil and palm kernel oil. There are also fractionation, hydrogenation, effluent, neutralizing, acidulation, packing and shortening plants. It's a huge responsibility to take care of a big plant.

I like

- It's enriching. I learn, practice and enjoy it. Lots of unit operations, process control and instrumentation subject. Some fluid mechanics and other subjects as well. I enjoy the learning curve.
Every day, I will always learn new stuff. Every day I will learn new interesting chemical engineering knowledge and practical experiences too.

- I can be the expert in certain fields. I'm gradually establishing myself in certain areas like heat exchangers/transfer, cooling tower, process control, control system, etc. There's a lot.

- Fantastic bonus and increment.

Dislike

- Sometimes there is too much job to be done. The responsibility is massive. Must be 24 hours alert of the production/operation (because I'm under production dept.)

- When something goes wrong in the plant, and the production target is not met, I have to answer my boss.
- I don't like too many meetings, either with my superior or supplier. A few meetings are OK.

- One problem is I live very far from my working place. It's basically a geographic problem. The distance in between is 50 km. Therefore, I've to travel at least about 100 km daily.

- Managing manpower/down line is another challenging task. This requires good communication and social skills. I'm still learning and improving myself in this area.

Bottom line, there'll be some pros and cons in our job. For the time being, I'll continue serving my present employee and I'll still enjoy exploring and learning.

"Manufacturing is more than just putting parts together. It's coming up with ideas, testing principles and perfecting the engineering as well as final assembly."

James Dyson
(Inventor, industrial design engineer & founder of the Dyson)

ISO and HACCP Audit

My plant was audited today and we thoroughly prepared for this occasion since last week. It was an ISO and Hazard Analysis and Critical Control Points (HACCP) audit. We have to ensure all records and documents are well kept and organized. We must clean the plant and clear it from unwanted objects such as cotton rags, welding rod, plastic, food wrap, cigarette bud, water clogging, etc. which were lying around (the lists of objects that you can find in the plant are too long but please don't assume we are not cleaning the plant. We clean the plant on a daily basis, but sometimes unwanted objects suddenly appear!!! It's a very big plant - 10 floors). Besides that the plate heat exchanger (PHE) and pump tray should be dry and clean (for your information, sometimes the PHE and pump leaks and oil is contained in the tray. This can be considered oil lost!!! Oil is money in the oil industry).

The Critical Control Point (CCP), [which is a point, step or procedure at which controls can be applied and a food safety hazard can be prevented, eliminated or reduced to acceptable (critical) levels] in my plant is the filter bags. The filter bag record and differential pressure was inspected randomly. This CCP is very important in food industry because it is the most critical step in ensuring the safety and cleanliness of a food product. I don't intend to describe more about this as it might be too technical (unless anybody wants to know more, you can ask me via email).

The glasses at the pressure gauge, temperature gauge and vessels were inspected as well. In addition, the pest control program was checked too. During the interview, we managed to answer and convince the auditor that the ISO and HACCP standards are well applied. I'm glad that we passed through this round of audit without any non-conformance (NC). Thanks to all the supervisors, shift leaders, plant operators and contract workers for their commitment, dedication and hard work for the audit preparation.

Chemical Engineers Come and Go

In a private company or organization, it is normal to see chemical engineers (or other engineers) come and go. They work for a duration of time and gain as much as they can, then leave the company. There's nothing much the company can do about the firm decision they made on leaving (bear in mind, this is a worldwide problem for employers). If the company considers the chemical engineer as an asset, they will normally counter offer and/or increase remunerations/benefits for the engineer. On the other hand, if the chemical engineer is not really performing, the employer will be more than welcome to receive their resignation letter.

I have seen few of my colleagues leave the company for various reasons. Just to share with all of you, various reasons that made them leave the company, among them are:

1. Bored!!! When an engineer has this particular reason to leave the company, he/she usually has already learned and experienced substantially a lot to the stage that he/she is in search of a new challenge. A friend of mine, having worked in a plant (downstream) for 8 years and established himself in his workplace, a multinational company, decided to leave. When asked why, he said he was already bored. He wanted a new challenge. He wanted to work in the upstream area. He managed to get a job in another multinational giant company, and I can see him very excited and looking forward to learn and venture a new field.

2. Pressure. This is a common reason for engineers to leave (but they would never say so). Sometimes, they are given huge responsibilities, tasks, projects, assignments etc. There'll be high expectation for them and they are expected to perform excellently. Not to mention numerous meetings they have to attend and reports to prepare. Various problems created by down-line and up-line levels may create some turbulence in the engineers' mind. He might not be happy working, constantly being pressured by his nagging boss. This can be a very strong factor for the engineer to leave. Frankly speaking, this is a popular reason many engineer quit their job. They simply cannot handle the pressure. They don't care about the money or the high salary anymore. I've seen a lot of these cases.

3. Opting for higher salary. This is quite an obvious reason - Money Matter... Money oriented..... Everything is all about money....Everybody has different perceptions on money. A colleague of mine, who is a senior executive, is offered a good high level position to manage a few plants, BUT, not in Malaysia, but instead in a foreign country. His salary was already quite high here in my company, but after he was offered quadrupled of what he was earning he could not resist. Well, on the other hand, I know he has a lot of debts to settle before retiring, and that's why he has to grab the opportunity. What about you?

4. Getting Promoted. When an ambitious engineer discovered that there's no chance of him getting promoted in the company he's working in, he might not waste a wonderful opportunity when another company offered him a higher position (which comes with a substantial salary increase, possibly with some profit sharing as well!!!). A friend of mine who's a dedicated engineer was offered to be a factory manager in a newly built utility plant. Despite of several huge amount counter offers (few thousands dollar) made by his current employer, he still insist to leave, even though his job at present is comfortable.

5. Wind of Change. Sometimes, an engineer is bored of the hectic technical life he lives in. He feels that he wants to experience a different working environment. He might consider giving up all his chemical engineering knowledge and skills to something with lower pace. I have a friend whom decided to quit and choose to become a teacher, which many consider far more relaxing than becoming an engineer. He is entitled to work half day and will also enjoy the school break (lots of holidays). At the same time, they can have one or two tuition session to earn something extra. Well, that's his choice.

6. Unsatisfied. In some cases, a chemical engineer might be unsatisfied with some internal political issues. Ideas and suggestions brought forward by the engineer may have always been taken lightly or simply pushed over. The engineer will have a feeling that the company is not appreciating him and his ideas/plans. This situation will trigger him to consider leaving the company. However, this particular reason is very rare.

7. Family matter. An example of this category is myself. I used to work in a local oil and gas servicing company. I enjoyed and loved my job as a project cum chemical engineer. Traveling and working, tasting upstream and downstream working environment really enriched me. Unfortunately, the company decided to move to Kuala Lumpur, the capital of Malaysia, in order to increase the chances of securing more oil and gas tender/projects. I was instructed to move as well. Unfortunately, I'm already settled in Johor Bahru (JB). My wife, who's pursuing her Ph.D. (chemical engineering), is doing very well as an academician, researcher and consultant. Furthermore, my parents are living in JB and we have also bought a house in a nice neighbourhood. So, I decided to hunt for a job nearby and here I am, working in a physical refinery plant, in JB (after resigning from that company, of course).

8. Combination of some of the above reasons. To be more transparent, I can say, most of the engineers leave because of a combination of reasons stated above. It's going to be lengthy if I explain or give examples here. But, the bottom line is, combination of the factors above will be a catalyst to increase the rate of reaction for the process of an engineer leaving and seeking for a better job.

My opinion/comment:
This is an interesting issue. It's the engineer's decision and everybody has to respect it no matter what reason(s) they have. Let's just hope that the move will be a positive one towards a better career and profession.

Next issue...
Nowadays, there are many graduate chemical engineers. Every year, thousands of chemical engineers are produced from various local and international universities. However, only small percentage of these chemical engineers managed to get themselves a decent chemical engineering job. This is a phenomena resulting from several reasons. I'll share the reasons why they failed to qualify and be employed in my next post.

Employers Expectation to a Chemical Engineer

In my previous post, I mentioned about how tremendous the amount of chemical engineering graduates is produced by universities all around the world every year, both locally and internationally. However, only small percentage of these chemical engineers get decent chemical engineering job. This is a phenomena resulting from several reasons. Why is this so?

1. Communication Skill

This is the prime reason why a chemical engineer usually doesn't get employed even with a superb chemical engineering degree/grade!!! There's no doubt that thousands of chemical engineers graduated every year. Some of them excel brilliantly in their academic. They score flying colours in all the chemical engineering subjects and many thought they will get good jobs. Unfortunately, when they are called for interviews and bombarded with series of questions, both simple and tricky ones, they froze! They have the brain but unfortunately they are not confident enough to convey the message verbally.

Most of the cliché initial interview questions will be "tell us about yourself". This is a vital question that needs good elaborations on who you are? If you fail to narrate your background in a pleasant articulate language, your future employer will get annoyed and be turned off. Your chances in acquiring yourself a job would be slim. You need to improve on the communication skills - how to communicate effectively, how to become a good listener, how to answer....and much more....You also need to have quite an extensive knowledge on the company you plan to work for to help you gain the confidence you really need.

An engineer is answerable to his superior or manager. At the same time, he has to manage his down line manpower which may include supervisors, superintendents, operators, technicians etc. An engineer also needs to know everything beyond his authority and power. One of the most effective ways is to get his down line reporting and informing on a daily basis. Sometimes even shift hour basis. The instructions must be delivered clearly to avoid silly expensive errors. Therefore, an excellent communication skill would really do you good in order for you to be a competitive engineer.

In my case, for example, I make sure I know every single occurrence in my plant and its present condition - 24/7. I ordered my supervisor to call and report to me at the end of every shift. This way, I will keep myself updated with the latest information without me being there to monitor the plant progress outside of my working hours. My supervisor will call me at 2230 hours and 0630 hours (after their shift) everyday regardless of where I am. Whenever any problem occurs, the supervisor will immediately inform me. That's why I have to bring my hand phone everywhere including into the toilet! With this, I'm well informed of all the activities and progress of the plants. I can seek for further details later but it is very important for me to be aware of the problem before my superior because I will definitely look incompetent if it were to happen the other way around. Believe me, the last thing you want is your superior to highlight and raise the problem before you do.

2. Analytical Skills

When I attended the interview at the company I'm working now, after giving the interviewer some background about myself, he began asking me some technical questions. He asked me to do some calculations. One of the test was to calculate the volume of chemical (corrosion inhibitor) injected into a pipeline (length and diameter are given) in such way that the chemical will stick onto the internal wall of the pipeline with 2 mil (1 mil = 1/1000 inch) thickness. He said, *"I'm going to the toilet now. When I return, I want the answer...."*. Can you imagine a question like that given to me? I have to force my brain to think and recall related formulas. Luckily, I answered it correctly, though I took some time to calculate.

At least, I proved him that I am capable of carrying out the chemical engineering calculations.

3. Degree Result

A lot of chemical engineering student would normally have the ambition to become an outstanding engineer. A very small percentage would opt to further their study specializing in certain niche area. Normally this tiny portion of students will become a lecturer or researcher, and they have no problem earning a master's degree or a Ph.D.

For me, my truly aspiring ambition is to become a lecturer; however, it is not easy to be one. We have to be a very excellent candidate and that is reflected from our result transcript/certificate. If we are lacking at that division, then we have to add more values to ourselves.

Some good universities really look on your certificate. Although I received few interview calls and offers from other universities, I have to reject them because they are located in other states in Malaysia. I have settled down and decided not to move (my wife is having a good career and my parent is also living nearby).

I worked very hard after my first degree to compensate the lack of quality to become a lecturer. I excelled in my Master's degree and I was hoping that I can use my master's degree as a ticket to get there. I have produced a number of papers in seminars and journals. Together with my research group we have won numerous awards. I've published bulletins and designed research posters (which my previous boss really liked). Unfortunately, at the end of the day, most of them still look at my degree instead of my true quality and the impact that I am capable to give. Well, that is life. Sometimes, we don't get what we want. God knows better. Luckily my wife becomes a chemical engineering lecturer. At least one of us managed to and I envy her.

Conclusions

Those are just three of the main reasons why an engineer might not get employed. There might be more, but the rest are not as important to me. I believe, if one can overcome all these three hurdles, one can be a good engineer and go further....

Note: The following reasons are from my personal opinion based on what I've been observing. It is not based on any influence from any source.

*"It is impossible to work in information technology
without also engaging in social engineering"*

Jaron Lanier
(American computer philosophy writer, computer scientist,
visual artist, and composer of classical music)

Making the Right Choice

When I was about to finish my 5th form (high school) back in 1993, I was forced to choose my profession in the future. The decision I was about to make will determine the course I will take for my first degree. Being in boarding school, we were always exposed to the courses available in the universities locally as well as in overseas. Given that much choices, I was wired up and confused of which way I have to go. It was so hard to make that big of a decision for yourself at that age. I got no help or whatsoever from my parents because they themselves are not well knowledged about the courses, what more the professions available. I live a completely different life from them, they were poor, and their parents were poor. They do not even get the opportunity to go to school in those days. So it was up to me.

What I know so far at that time was I like chemistry so much. Just give me any chemistry exam question and I will know how to answer them. All the chemical reactions were in my head and the calculations were so easy for me. Then as I look through the courses, the choices were narrowed down to two: chemistry and chemical engineering. Being a chemist did not sound as exciting as being a chemical engineer, so I thought. And the rest is history.

The first few years in chemical engineering course were really a struggle to me. It was so hard to accept the fact that this course does not involve so much chemistry but instead physics, a subject that is not really my cup of tea. I cursed myself, I didn't want to study and I ended up being in the 'under-probation-about-to-be-expelled' list. My grade was so bad during my A-level and I broke my parents' hearts when they received the letter. That was an eye opener for me and from then I decided to not grieve over the choice I've made and prevail through it. I started to make notes in classes, did my assignments and attend classes without failing. I was so proud when I got the highest mark in my class and appointed as a tutor for weak students in college.

It is very important for you to choose the right course or profession because it will affect your life. Make sure you have every bit of information on the courses and prepare for it mentally and physically. This blog for example, is not only a good blog for chemical engineers and chemical engineering university students but also an excellent tool for students at high schools to expose themselves about chemical engineering.

My personal apparition on what chemical engineering is about is how to realize the experiments being carried out by chemists in labs in order to produce a chemical product/chemical invention. The 'how' includes equipment design, process control and plant operation among all.

- Zura

"Engineering stimulates the mind"

Bruce Dickinson
(English singer, songwriter, musician, airline pilot,
entrepreneur, author and broadcaster)

Wrapped up

Maybe many of you do not realize that the last post was written by me, Zura, Zaki's wife. Well, the 'posted by Zura' is written so small at the bottom of the post anyway. I forgot to mention that it was me instead of Zaki in the post itself. He invited me as a contributor or a guess blogger to his blog.

As for now, Zaki is too wrapped up in managing his plant and is mentally tortured to prepare his monthly plant report. This is his first as he is just assigned to be the Head of Department recently. His time in the office flies by and it is very restraining to his eyes from constant staring at the computer monitor. When he comes back home, he is literally drained and it is very difficult for him to spend the evening in front of the laptop at home. I hope he will pass this stage of his career and soon can blog normally again.

Well, that's the reality of being a true chemical engineer. I wouldn't really understand wholeheartedly because I have never had any experience working in the industry and I found it quite hard to teach something that I've never done. If it is not because of the rarely held plant visit for the faculty staffs, I wouldn't even know how a real distillation column looks like. It's ironic isn't it? Luckily we, the lecturers, just have to teach the theory behind designing those equipment's in chemical plants. To really take care of a plant, you will have to experience it yourself and face all those troubleshooting task that you chemical engineers have to carry out in order to make the plant run smoothly.

Here in the faculty (FKKKSA, UTM) when I was an 'active' staff, meaning that I was not on study leave, I have to teach a subject called 'Unit Operation I and II' (now named Separation Processes 1 and 2). This is a subject where you have to understand all the unit operations or equipment involved to separate gas-liquid, gas-gas, gas-solid, solid-solid, solid-liquid, and liquid-liquid mixtures of material in a chemical plant. This is including distillation columns, absorption columns, evaporators and crystallizers, among all. It is a tough subject for the students because the group of lecturers teaching this subject normally gives exam questions which are out of the world difficult and unthinkable. The students fear this subject and the exam is full of suspense. As for me, I try to be as moderate as I can, but I certainly want to test the students' understanding on the subject.

I miss teaching but now I have to concentrate on my PhD so that I can serve my faculty again and give my little contribution in moulding the national assets.

- Zura

*"I don't spend my time pontificating about high-concept things;
I spend my time solving engineering and manufacturing problems"*

Elon Musk
(Business magnate, investor and engineer, founder, CEO, and lead
designer of SpaceX; co-founder, CEO, and product architect
of Tesla, Inc.; and co-founder and CEO of Neuralink.)

You Should Love Training and Seminars

I love attending training. I don't care what type of training because I can learn new things and skills. Since I work, I have attended various in-house and outside training as well as other free and paid seminars.

Among the things/new knowledge's that I have learned so far are:

- Bleaching earth - chemical composition of the silica alumina solid bleacher...
- Cooling tower/boiler chemical treatment - technique, monitoring, dosing etc...
- Instrumentation/flow meter for plant - Various flow meter principles...
- Preventive maintenance - Need to organize this to avoid expensive downtime....
- Air compressor operation - They have a better improved compressor....
- Chiller operation - oooh... They use that type of Freon as refrigerant...
- Shortening plant - Now I know how they make margarine...
- Load cell - Never knew about the technology behind weighing various objects....
- ISO and HACCP training - Now, I'm an internal auditor...
- Lean-Kanban - I learn to improve my working system...
- TQM - There's a lot of other quality system in TQM....
- GC - I learn how to operate a gas chromatography hands on
- Helicopter Underwater Escape Training (HUET) - physically challenging training for going offshore...

Well those are just few of hundreds of formal and informal training I have been through. That's why training/seminar is important to develop our skills and knowledge.

Another good thing that I like when attending training/seminar: "We get souvenirs". So far, I have collected a lot of gifts and souvenirs such as pens, t-shirts, note books, key chains, cap, diaries, calendars, mugs, and many more. It's good to get free stuff....isn't it?

Bleaching Earth Bag Support

Every day is a different day at work in terms of the knowledge I acquired. Learning new things on a daily basis gets me excited. It's not necessarily just to learn on high tech theoretical knowledge like heat transfer between plate heat exchanger or the reaction inside the distillation column. Sometimes, I learn simple stuffs too and that makes me say...."Ooooo... I never thought it was like that....." or ..."That's interesting, why haven't I paid attention on that....".

The past week, I have several assignments to do. One of them is to sketch/draw/design a rack to support a 850kg bleaching earth bag in order to convey them into the hopper on top of the plant. We don't use bleaching earth bag all the time. We just keep them for back up. However, after a period of time (month++), we need to consume them before the bleaching earth activity and quality deteriorate. Currently, we use a forklift to lift the bag and pour the bleaching earth from the bottom of the bag into a small hopper. The process takes 2-3 hours. Within those few hours, the forklift is only occupied for one job i.e. to support the bleaching earth bag and it could not be used for other jobs. To avoid this from repeating, a rack can act as a support for the bleaching earth bag. I made a drawing of the rack. However, in our maintenance meeting, my colleague suggested to use a metal pallet (for packing small drums) as a rack by positioning it upside down. Well, that's not a bad idea at all. I never thought or realized we have metal pellets even though I walk pass through the packing plant every day.

Maybe some of you are wondering what is bleaching earth?

Bleaching earth is a highly effective adsorbent essential in the refining process for the purification and decolourisation of edible and non-edible oils, fats, tallow, petroleum jelly, paraffin oil and waxes.

Bleaching earths are special clays activated by physical or chemical processes. Natural clays are found in special strata and are subjected to various processes such as desegregation in water, treatment by sulphuric acid solutions, filtration, baking, grinding, etc. Efficiency of bleaching earth depends on selecting the right grade and proper blends of the basic raw material, bentonite obtained from various mines.

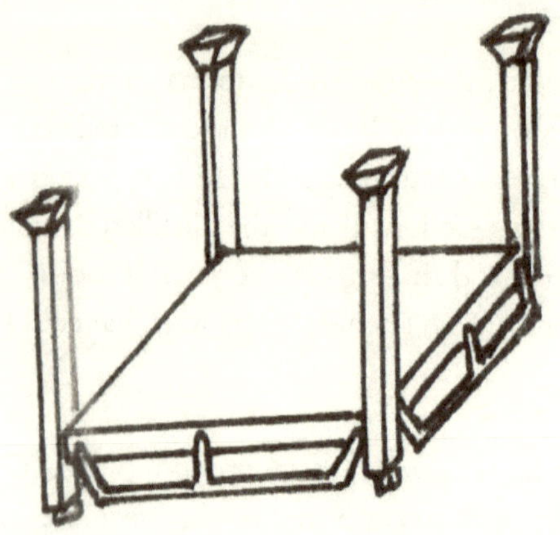

What we have to do is just put the metal pallet upside down, and cut a circular hole on the centre. Then we can place the bleaching earth bag on top of it with the small hopper underneath the upside down metal plate. The small hopper will then convey the bleaching earth to the plant to mix with oil.

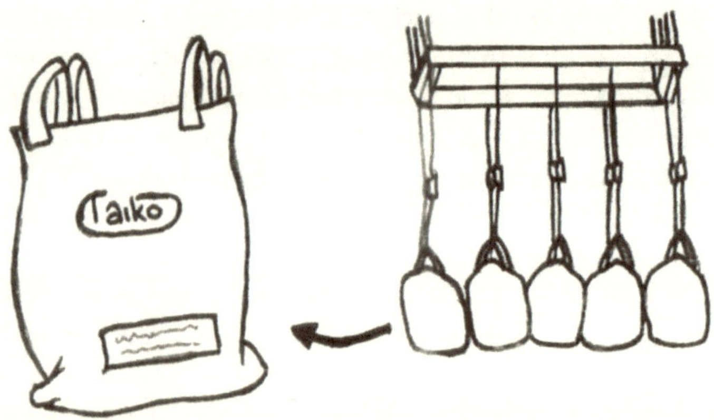

Example of the bags that we'll place on top of the upside down metal pallet. There's an opening at the bottom of the bag to pour the bleaching earth. The opening and bleaching earth flow can be throttled by using a rope which comes together with the bag's bottom opening.

Association Memberships

I'm a member of few associations (as of 2007). To name them:

-IChemE (Institution of Chemical Engineers, UK) - Associate Member
-MOSTA (Malaysian Oil Scientist and Technologist Associations) - Member
-UKAN (United Kingdom Alumni Network) - Member
-Bradford University Alumni - Member
-Universiti Teknologi Malaysia Alumni - Member
-BEM (Board of Engineer, Malaysia) - I'm working my way to be a professional engineer

To tell you the truth, I'm not an active member of those associations mainly because I'm too fully occupied with my chemical/process engineering job. It's not easy for me to take leave and attend annual grand meeting or conferences etc. I'm not as flexible as my wife, who is a chemical engineering lecturer/researcher/consultant. However, I keep track on the activities and progress of those associations. It's good to attach ourselves with professional associations because we are constantly updated with current news and events. This is very useful to establish our career and networking.

IChemE, MOSTA and BEM membership require some annual fee. For IChemE, luckily last year, they managed to form a local branch in Malaysia and the fees are charged in Ringgit Malaysia currency instead of Pound Sterling (which is more expensive for us....). I have not renewed my membership for IChemE and MOSTA this year. I need to do so as soon as possible.

Guess What This Is?

Slowly guess what the following items. If you scroll quickly, you'll miss the suspense in guessing what the equipment is.

a

b

c

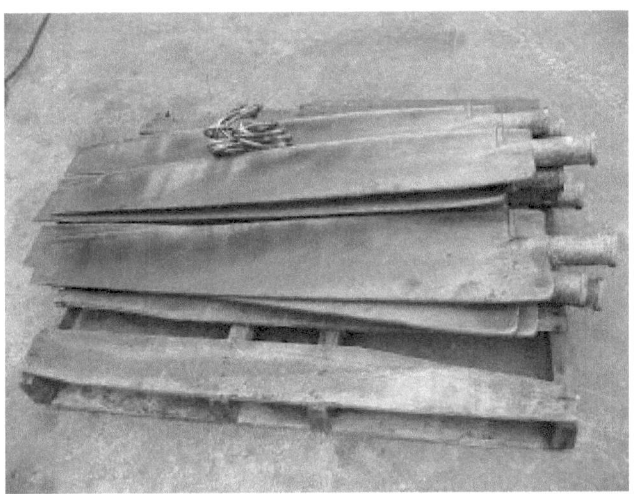

d

e

f

a: Motor (for cooling tower fan) - the bearing has broken. It's vibrating.
b: Blade holder removed from the motor using hydraulic clamp
c: Blade holder
d: These are the blades
e: Looks like a fan
f: It's a cooling tower fan...

Did you guess correctly?? Those are all the cooling tower fan components.

My Production and Process Engineering Duties

It's May 2007 and I have to prepare the monthly production report for April as fast as possible. I don't want to delay them anymore and make others' suffer waiting. I'm trying my best to work smart and systematic.

Today, I was presented with a practical student (chemical engineering course) from a local university for me to guide. He'll be around for the next 9 weeks. I'll be exposing him to the plant processes, processing parameters (steam, electricity, water, air, bleaching earth etc), supporting units (such as boiler, cooling tower, heat exchanger etc), oil quality monitoring and testing and much more. I introduced him to the morning shift and gave him a short basic tour in the plant. Then I left him in the control room for him to interact with the supervisor and operators (having some informal learning process). I also gave some 15 minutes short basic on-site practical explanation on various types of valves (ball valve, butterfly valve, globe valve etc).

My manpower is another area that really concerns me. Unacceptable disciplinary problems from few operators made me feel very irritated! I don't understand why they treat their rice bowl lightly?!! I have given some short verbal talk and advice to related person. If the problem continues, I shall have to introduce harsh action, possibly a "warning letter" or "domestic inquiry". I hate to do this, but this is really part of my job as a process engineer. As engineer we do not just take care of those technical and report stuff, but social - manpower issues as well.

Some of them expressed their dissatisfaction after receiving their bonus and annual salary appraisal letter. Some of them smiled. I can guarantee those who diligently work and maintain excellent disciplinary record a good increment next year. I'll do my best to fight for those who deserve.

Some Updates from an Engineer

The past few days were very busy days for me. However, I tried to get a few minutes off from my hectic activities (event at night! after work...) to post something here. Let me just share few things that I did today and yesterday at work (and after work) not in particular order:

- Watching, witnessing, and experiencing civil, construction job performed at my work place. I learned interesting stuff about how the civil work to expand, upgrade and modify a plant is made. Instead of normal irritating noisy piling, I saw quite smooth hydraulic piling and the machinery involved in running the operation. It's impressive.

- Meeting with suppliers. Today and yesterday, I received a visit from two different suppliers from the same niche area (both of them are competitors...). Usually, we don't really entertain suppliers, but both that came were worth the time. We learned and discussed in order to improve our work. We exchanged information and ideas as well as shared experiences. It was a fruitful discussion.

- My cooling tower fan was dismantled because it vibrates. A 20 ton crane was used to lift the 7 bladed fans down to the maintenance workshop. It's a new experience looking and inspecting the fan closely. This is what we call "Preventive Maintenance".

- Supervising practical student. This is another area that I'm covering as well. I made a format/guideline for him to study, research and investigate. He work his way according to my guideline. Every end of the day, we shall sit together and discuss his work/progress. In order to develop his communication skill and improve his second language, I asked him to converse with me in English. I hope he will learn as much as possible within these 2 months practical training period.

- New Heat Exchanger. I ordered a new set of heat exchanger and it would arrive from German in two months' time. I'm checking on the perfect location to install the heat exchanger. I have to consider the piping work (civil work) and flow of oil. Parts and materials for the heat exchanger must also be finalized such as ball valve, temperature gauge, pressure gauge, pipes, expansion bellow, insulation, RTD, mass flow meters and others....It's challenging, interesting and fun....

- Managing/thinking about my manpower problems. This is a real head-ache. No wonder there's a saying, "It's easier to handle 100 cows compared to handle 10 peoples". Each of them has different behaviour and attitudes. However, only very few of them continuously have disciplinary problems. The others are good. I'm calling for a meeting tomorrow to discuss and find a way to eliminate social and disciplinary problems within my down line staff.

- Report preparation. This is a daily routine activity. However, I have not yet completed the monthly report because I'm still waiting for the costing report from accounting departments.

- Supervising design project. I'm also supervising a final year design project from a local university (few months back, my friend (a lecturer) asked me a favour to be an allied supervisor from industry....). I received the report just now and I'm checking it at home (at night after work). I'm going from line to line and page by page. I'm impressed with the report. However, there are still few errors here and there. I just reached page 37 out of 78. There are 41 pages to go through.

- I received an email from a reader of this blog (you know who you are... :). He's currently working on his final year design project entitled "The production a hydrocarbon based methanol". I don't have much information on this. However, I'll try and check it out. I'll get back to you as soon as I get some valuable information related to the topic.
Well...there's more....but, I need to take a break and continue reading/checking the final year design report.

Danfoss Inverter - Save Your Power

I attended an interesting training in July 2007. It was a training about inverter conducted by Danfoss. Danfoss is a big company based in Denmark producing various products. The trainer (product manager) was really well prepared. We were given a file complete with colourful brochures and slides as well as papers to jot down note. The introduction of the presentation was very well defined. The presenter informed us the flow of presentation together with the expected time (in minutes) for each topic. All together 120 minutes (2 hours). That's good. At least we know when the training will end.

The training commenced with a short 5 minute video introducing Danfoss. It was a very impressive and informative presentation. They really have state of the art technology (which utilizes efficient and effective robots) to manufacture inverters as well as other products.

Inverter can save power and money (reduces maximum power demand). In addition to that, it can lengthen the life span of your motor and other equipment. It reduces start current and reduces start torque by the square of the current reduction. Inverter actually reduces voltage starting attempts in a motor and avoids it to start instantaneously. Inverter will slowly and gradually supply the voltage and current to a motor until a maximum level.

There are more to discuss about inverters. Inverters are really impressive in saving power. A lot of process plant are installing and utilizing inverters

Vacuum Drop Alert!!!

The entire week has been interesting. Besides my routine job, I learned and experienced new things.

Vacuum dropped due to wet steam

I hate this case!!! When vacuum pressure dropped, the quality of oil is affected and most of the time the oil have to be rejected. In this case, vacuum drop occurred because of wet steam. Wet steam is basically water carry over from the boiler. Boiler is supposed to produce steam and deliver it to plant. When boiler began sending steam + water to the plant (which is not supposed to happen), the vacuum system is interrupted. The vacuum lost its capability to suck fatty acid which is supposed to be removed from the oil. I could not let it happen again. I don't want my plant to reject oil again because of this 'typical' reason. The boiler people never want to admit that they sent steam + water to the plant.

Therefore, I decided to investigate the case further. Few minutes after the 3rd incident occurred (wet steam) I took a bottle of condensate water sample (this is the condensed steam which became hot water) and tested its TDS (total dissolved solid). We tested the TDS using a digital TDS meter at the boiler house. The TDS of the sample showed 70 ppm. Normal condensate water should display reading less than 10 ppm. This implied that water carry over (wet steam) happened. Why?

Because, steam is water (H_2O) at 120 ++°C temperature (around 15-17 bar). Hence the boiler is/should only send steam (pure H_2O). Pure H_2O should have zero or very minimum content of metal/solids/substance inside it. When steam condensed and became condensate water, the content of metal/solids/substance should remain. However, when wet steam occurs (due to water level fluctuation inside the boiler); some water follows the steam to the plant. The water is chemically treated and this increased its TDS. Therefore, if the condensed water is 'rich' of TDS, this confirms that wet steam is taking place.

Up till today, I gathered 2 condensate water samples (as proof) after the wet steam incident and the TDS result is very high!!! The boiler people definitely have to do something to avoid this from worsening or happening again. I'm not accusing anybody to be responsible of the "oil rejection". I sincerely just want to solve the problem. Save cost and time as well.

Checking the Total Dissolved Solid (TDS) at different temperature

When I first checked the condensate water sample which was collected by my supervisor, it showed TDS of 67 ppm. The boiler man said that is impossible. He claimed that my sample was very cold (I left the sample in my office for about 5 hours before testing it. Therefore the air condition has cooled down the condensate water temperature to about 24-27°C). That's fine....I told him, let's heat the same sample up to 65-70°C and repeat the TDS test. We heated up the sample and re-tested the sample at 65°C. The sample showed reading around 70ppm. At least, it is close to the earlier TDS reading. Before repeating the TDS test, I told the boiler man that TDS have nothing to do with temperature. However, he stubbornly claimed that temperature influenced the TDS reading. After that he kept quiet and admitted that temperature have nothing to do with TDS. Hence, the condensate water really has high TDS which means water carry over (wet steam).

Getting a Hoist

I need a hoist in the store (warehouse) to lift pallets and other heavy stuffs efficiently. Therefore, I looked and searched for companies selling hoist. After calling and requesting quotations from few companies (selling hoist), I browsed the net to have a sneak peak on those companies to see what are the products and services they have to offer. It was very interesting to see that there are various types of hoist with various applications and various funny names too. I never knew that hoist have a lot of "species!!" I don't have much time to go through all of the hoist information, but I'm glad I learned few of them.

The Wikipedia defined hoist as "a device used for lifting or lowering a load by means of a drum/barrel around which rope or chain wraps. May be manually operated, or electrically or pneumatically driven and may use chain, or fiber or wire rope as its lifting medium." (en.wikipedia.org/wiki/Hoist)

Among companies selling hoist are harringtonhoists.com (USA), hoistlift.com (USA), www.grainger.com and mstime.com.my (Malaysia). There are also lots of interesting names for their hoists such as "Street, Donati, Italkrane, JD neuhaus, MS Herkules and Daesar."

Very interesting too there is a magazine "specializing in just hoist".
Check out hoistmagazine.com.

Cooling Tower FRP Water Distributor

What you can see in the photo is a cooling tower deck. This is where warm cooling water from processing plant arrives. The warm cooling water will then be evenly distributed on the light blue colour FRP (Fiber Reinforced Plastic) deck by a water distributor. The warm cooling water will then falls underneath the deck through the target nozzle holes. You can see there are plenty of small black round shape holes on the deck. The target nozzles are important to sprinkle the water before it goes further down to increase its surface area (I'll show and explain about target nozzle in future post). The warm cooling water will then be sprayed over a fill in the cooling tower to increase the contact area, and air is blown through the fill. Majority of heat removed from the warm cooling water is due to evaporation. The remaining cooled water drops into a collection basin and is recirculated to the plant (for chiller or heat exchanger). Typically, the temperature drop is 10°C.

Cooling Tower Target Nozzle

As promised, I'll share some information about target nozzle which is among vital parts of a cooling tower. A target nozzle looks like the photo shown below. That is a "Counterflow Nozzle Orifice" (sizes available from 1/2" - 1 1/2). A target nozzles function is to sprinkle the warm cooling water to increase its surface area (for heat transfer with air) before it goes down from the deck to the water basin. It is usually an injection molded polypropylene unit consisting of two parts—the main body with integral target diffuser and a snap-on insert or orifice cap. A cooling tower deck may have up to 200 or more target nozzles.

All this while I only know this type of target nozzle. After some research, I found there are various other types of target nozzles as shown below:

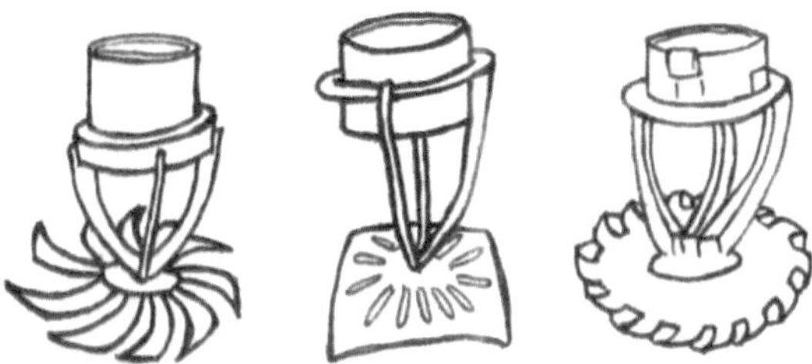

The first type (left) is a Counterflow Nozzle - 2" NPT Thread, (Orifice sizes 1/2" - 2"); the second (mid) is the a counter-flow down-spray nozzle; Square-Pattern, Full Coverage Type, 1-1/2" Standard Pipe Thread Connection; and the third (right) is a counter-flow up-spray Nozzle, fits 1-13/16" Diameter Hole in PVC pipe. I began to appreciate all components in a cooling tower.

How Do You Analyse Your Heat Exchanger Performance?

I have more than 10 plate heat exchangers (PHE) in my plant. Usually, we'll monitor the flowrate, inlet and outlet temperature, as well as pressure. Before, I joined the company; I observed that the "cleaning in place" (CIP) using caustic solution was carried out when the PHE could no longer produce the desired flow rate and temperature outlet. This is one method of eliminating the fouling and scaling on the plate surface in order to improve the heat transfer. If CIP does not work, we have to dismantle the PHE, clean the plates in hot caustic and attach new gasket on it. This will be a far more expensive option and take longer time.

To avoid massive fouling and scaling from getting worse, we have to conduct CIP regularly. When is the right time to perform CIP? As mentioned above, when the flow rate is low or desired outlet temperature could not be achieved, we shall consider doing CIP.

However, I came out with a formula to calculate the "Overall Heat Transfer Coefficient" (U-value) of the heat exchanger. The U-value will give me some indication on which heat exchanger is providing the worse heat transfer. By simplifying the formula in excel file, I can easily know which heat exchanger needs attention just by inserting the inlet/outlet temperature, mass flow rate, heat capacity of oil, and surface area of the plates. The LMTD (log mean temperature difference) will be obtained from the formula and further applied to get the U-value.

I use and manipulated the following formulas:

$$Q = mc\theta = UA(LMTD)$$

Can you work out the formula?

Preparation Work for New Plate Heat Exchanger

In early 2008, I ordered a set of new plate heat exchanger (Schmidt-Bretten brand). We desperately need another plate heat exchanger (PHE) as a back-up unit to those that we presently have in plant. I have to correctly specify the specifications of the PHE as it is not a cheap piece of equipment. Thank God, everything was OK and we'll be expecting the PHE this coming August.

However, the job does not end there. I have to plan and prepare for the piping layout and other related civil work. I need to order additional parts/equipment's/instruments such as globe valves, ball valves, drain valves, control valves, pipes, RTD's, pressure gauges, vortex flow meters, temperature gauges, steel plates, bellow seals, bolts and nuts etc. (those items are very expensive!!! especially the globe valve...). We have to get all the piping ready before the PHE arrive. We need to complete it in less than 2 months from now. I pray and hope everything will run smooth for this small project of mine.

"A good scientist is a person with original ideas. A good engineer is a person who makes a design that works with as few ideas as possible. There are no prima donnas in engineering."

Freeman Dyson
(English-born American theoretical physicist and mathematician)

Spiral Heat Exchanger

In my plant, there are two small spiral heat exchangers which function as a plant shutdown cooler. I'm not sure what is the capacity and specifications of it. I have no experience using or operating them. But, I'm interested to learn about them.

Spiral heat exchanger differs a lot from plate heat exchanger (PHE) and shell and tube heat exchanger (S&T). There are also pros and cons using spiral heat exchanger compared to PHE & S&T. From my limited experience, a spiral heat exchanger requires less maintenance. There are no plates and no internal gaskets. To clean the internal part, a "cleaning in place" (CIP) is carried out using caustic solution up 60-75ºC. The spiral heat exchanger is totally stainless steel solid and it is very heavy. One problem that can make a spiral heat exchanger useless (or very expensive to repair) is when the internal wall crack or leak. When this happens, fluids entering the spiral heat exchanger will be mixed and hence affect the production quality (reject!!!).

Spiral heat exchanger exhibits ideal heat transfer and fluid handling characteristics for a wide range of applications. It takes up only one sixth of the space which means lower costs for buildings, pumps valves and piping
It uses 75% less pumping energy, provides higher K-value and a close temperature approach. Nowadays, there are very few companies selling spiral heat exchanger compared to PHE and S&T.

Example of a big spiral heat exchanger. The mass of it is about 5 tonnes.

Some Updates

I was very busy for at this time of the year (it was 2008). I was not updating my blog frequently. Nevertheless, as usual, I experienced a lot of new things at work. Let me share some of them with you.

<u>Between working in the office and at plant-site</u>

I have been an office dweller since a few months back because I was busy doing paper works, reports, planning, documents/filing 5S, meetings, appointments with supplier, trainings etc (Once in a while I came out and toured the plant, visited the control room, discussed few things with the supervisors and plant operators). After some time warming up the chair in my cold well air conditioned office, I felt really bored sitting in the office staring at the 15" LCD monitor. The past 3 days, I spent majority of my time in the plant and had my hands dirty. I equipped myself with a meter tape and pens of different coloured ink. Normally, when I walked in the plant, I'll be sweating. Some section in the plant is really hot (like a sauna!). However, the past few days were luckily colder (thanks to the rainy days) - So I sweated only a little! We (myself and my senior colleague) planned to install an alternative back-up pipeline for some section in the plant where tendency of blockage is most likely to occur. I have to come out with a plan on which route and how to install the pipeline. With that, I also have to calculate how many valves, elbows, flanges incorporated with the new pipeline.

In addition to that, we're going to install a new plate heat exchanger (PHE). The PHE will arrive somewhere in August. Other materials for the PHE such as piping, valves, elbows, expansion bellows, sockets, metal plates etc have arrived. I have measured and marked the location/position for the PHE installation. It's not a big upgrading project, but surely I learned and experienced a lot throughout the process of getting and installing a new PHE. I hope after the installation of this heat exchanger, our heat profile can be improved and the natural gas consumption can be reduced (utility saving means production cost saving).

One of our plate heat exchanger has undergone cleaning in place (CIP) for 2 weeks already. We consumed a lot of caustic and decarbonizer in order to thoroughly clean the plate heat exchanger. Yesterday, we stopped the CIP, flushed the pipeline and heat exchanger with water and air. Then we began using the heat exchanger. Today, I checked the flow rate and outlet temperature. Unfortunately, the temperature is not as high as it is supposed to be. We checked the outlet temperature from the control room (The RTD {resistance temperature detector} installed at the heat exchanger oil outlet sent signal to the PLC and therefore we can see the temperature reading from the Human Machine Interface (HMI)/monitor/PC in the control room).

I was not satisfied with the situation. We consumed a lot of chemicals, water, time, man hour etc to perform this CIP. The temperature should not be like that. It should be good. I personally inspected the pipeline and heat exchanger. My intention was to double check the temperature with the temperature gauges installed at the pipeline. There were all together 3 units of temperature gauges that can be referred and the temperature readings were all higher than the RTD reading. I knew it.... Earlier, I already suspected that the RTD is not right. The RTD needs to be serviced and calibrated. From here, I learned: Never 100% trust your instrument. They can be a good indicator for process control and instrumentation, but we also need alternative/back-up equipment/instrument to counter check the readings. Sometimes we can be deceived and make expensive wrong decision just from a false instrument reading. This will cost money and time which is very precious in this type of industry.

Manpower Issue

I received a resignation letter from one of my sharp and skillful plant operators. I have high hopes and plan to promote him to a higher position next year. Unfortunately, this is a global and universal issue that all private companies encountered. After chatting with him, I realized he has multiple reasons for leaving the company after serving for 9 years. Hence, I accepted his resignation and respected his decision. I just wish the very best of luck for him and hope he will be successful in future.

With his departure, I have to recruit a new plant operator and train him. The criteria that we require are very simple - (1) A disciplined person which means, will always come on time for work and would not be absent. (2) Willing to learn and work hard. (3) Take care of the cleanliness and smoothness of the plant.

I think those requirement should be ample. We don't need good academic result plant operator. Their attitude is the most important trait. With good attitude, they can climb the ladder of success. Just like my previous senior production executive colleague who started just as a cleaner/sweeper (25 years ago) but ended up now very successful as a factory manager taking care of a number of plants and projects in a foreign country.

Wilden Pump Seminar

It was in July 2007 that I attended a half day Wilden Diaphragm Pump Seminar organized by Winston Engineering in Hyatt Hotel Johor Bahru. I like going to this type of seminar or training held in good 5 star hotels. Besides learning about the products and technologies, we were served with very nice yummy foods. It was a very informative and enriching seminar. Sometimes we can meet with colleagues from other companies and do some networking as well. OK, before I proceed, let me provide some basic information or definition of a diaphragm pump.

"Air operated pump that uses a flexible diaphragm to separate the pumping chambers. Handles high viscosity liquids or liquids with suspended solids" (definition from bascousa.com).

"A diaphragm pump is a positive displacement pump that uses a combination of the reciprocating action of a rubber or teflon diaphragm and suitable non-return check valves to pump a fluid. Sometimes this type of pump is also called membrane pump" (definition from en.wikipedia.org).

Inside a diaphragm pump which are from various sizes and materials
(rubber, plastic, Teflon® PTFE etc....)

Various types and sizes of diaphragm pumps which are made from stainless steel, aluminium and alloy. There are also two types of joint: bolted (stronger, tougher and can withstand higher pressure) and clamp (easy maintenance)

This is the demonstration set for the Wilden Pro-Flo X which is their latest diaphragm pump. The pump is capable to deliver similar flow with lesser air consumption which means we can save on our utility cost - the air. The comparison is made with the older version of Wilden's diaphragm pump.

A little bit something about Mr Wilden...

It all started from a person named Jim Wilden who was working with a company manufacturing steel/metal back in the 1950s. He always faced a problem while pumping using an electric pump because his working area is always wet and can be dangerous with life electricity around. At the same time, there is plenty of air (compressed air) around his work place. So, he kept thinking on how to overcome the problem he is facing. Finally he invented the first air operated diaphragm pump. I hope that story is correct, because that was informed to us by Wilden's Asia Pacific Manager during the presentation introduction.

Jim Wilden pioneered a revolution, inspired a generation of innovation and created an entire industry. Wilden Pump & Engineering Company was founded in 1955 on the sketches of a unique but reliable, utilitarian pump, the air operated double-diaphragm pump. Today corporate ownership and "stand alone" local management provide Wilden with the clarity and focus needed to produce quality products Wilden products are engineered to meet today's industrial and commercial demands.

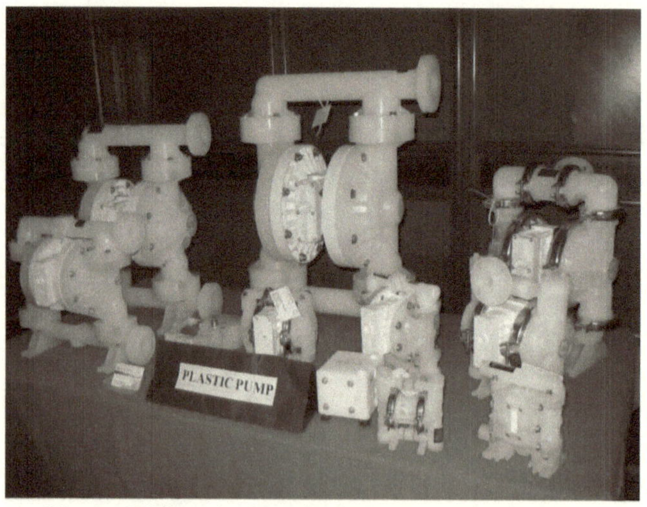

Oh yea...Wilden also have diaphragm pump made of plastic.
I'm not sure this pump is suitable for what industry?

This is Wilden's latest patented pump which can run without pumping fluid. Normally, pump will be damaged if they run empty, but this pump is not like that. Can you see the ping pong ball floating on top of the discharge line of the pump?

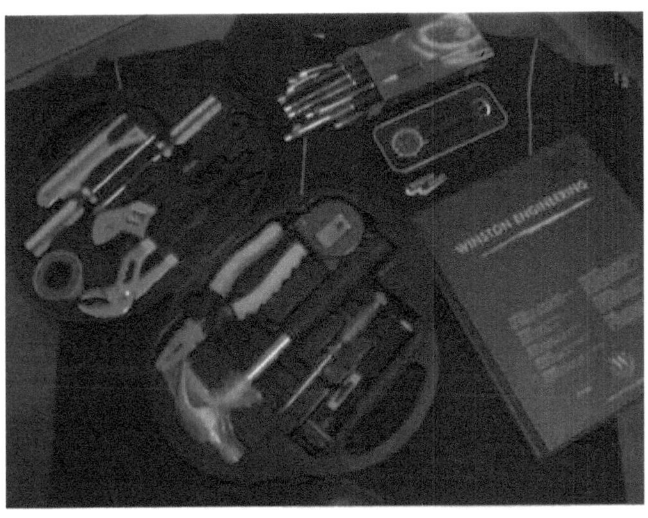

These are the lucky draw prizes that were awarded to me. I got a nice compact mini toolbox, a silver wrist watch, a name card book holder, and 12 pieces of ball pens. As a participants, I also brought back a jersey type t-shirt (I got extra red t-shirt - they gave away another t-shirt for those 25 participants who came earliest) and a 3 ink coloured ball pen. I managed to request another set of bags come with goodies and freebies for my boss. It's really fun getting souvenirs like this ☺

Bypass Line Configuration

There will always be a bypass line in a process plant. However, are you having the correct bypass line set-up? A bypass line set-up is different if the line is used for steam or oil or slurry. If you are using the bypass for a steam line, you can have the bypass line at the bottom of the main line. You must also have a steam trap installed on that bypass line to ensure condensate water is released when necessary.

What if you have a slurry line?

The bypass lines should be placed above the control valve (as shown in the photo) so that the slurry cannot settle out and build up in the line during bypass. In addition to that, slurry lines should be sloped 1/2" for every 10 feet of horizontal pipe to avoid settling. Actually the steam line should have certain degree of slope to hinder water from settling and further create water hammering which can destroy the pipeline.

That's how crucial the position of a bypass line is. We cannot simply place it in any configuration of position as we like. There will always be a reason for doing it.

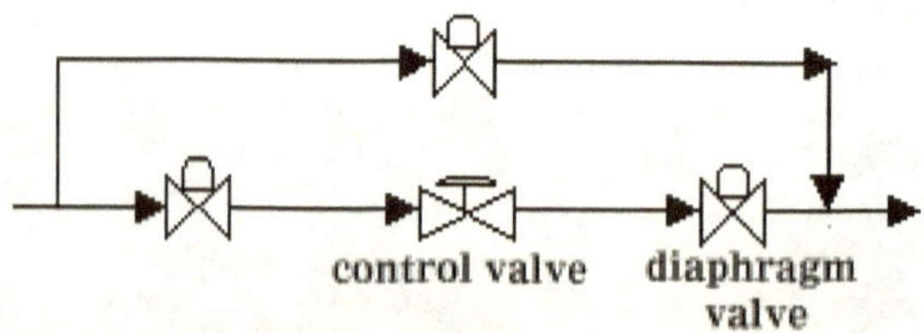

PIPOC 2007 KLCC

I went to PIPOC (International Palm Oil Congress) 2007 held at Kuala Lumpur Convention Center (KLCC). It was a day trip (we (3 engineers) travelled all the way from JB) and I just returned home, exhausted. The exhibition was not as big as I expected. Last year's OFIC 2006 (Oil and Fats International Conference) was better than PIPOC 2007 in terms of company participation. It took me just above 1 hour to complete touring the entire booths.

I did not attend the technical paper presentation this time. I just visited the exhibition only.

Various types of mini palm oil fruits. This palm oil gene produces a shorter palm oil tree for easier fruit plucking.

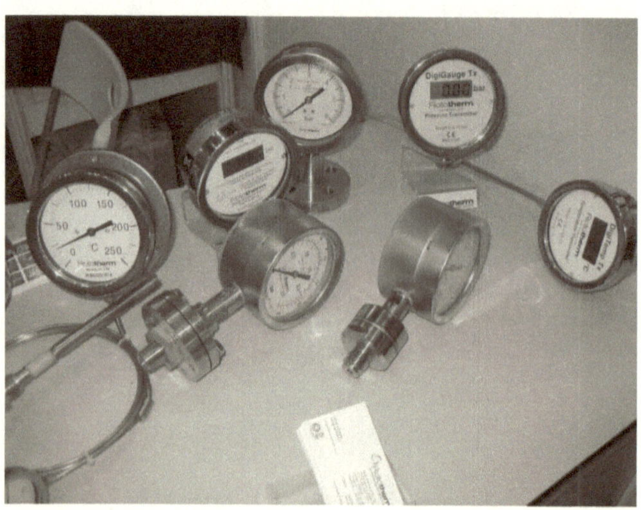

Various types of digital pressure gauges. I only use analogue pressure gauges and pressure transmitter in my plant. Digital pressure gauge reading is more accurate and is easy to calibrate on site

Butterfly valves of different sizes.

Horizontal leaf filter

More from **PIPOC 2007**

A little bit more photos captured from PIPOC 2007 exhibitions in Kuala Lumpur Convention Centre for me to share with you. Hmmm....Just to let you know, I'm not being paid for reviewing the products. I just took some photos and share them here. I hope you can benefit from it and in case you need more information, you can go to their respective websites to explore more about the products and its features.

A bellow seal stop valve - ARI FABA maintenance free bellow seal stop valve is designed with double wall bellow seal as standard. The sizes range from DN 15 to DN 400 (1/2" to 16"). Connections available: Flanged end to PN16, PN 25, PN40, ANSI B16.5 Class 150 and Class 300; Butt weld end and socket well end. I took this photo from Valmatic Engineering Sdn Bhd.

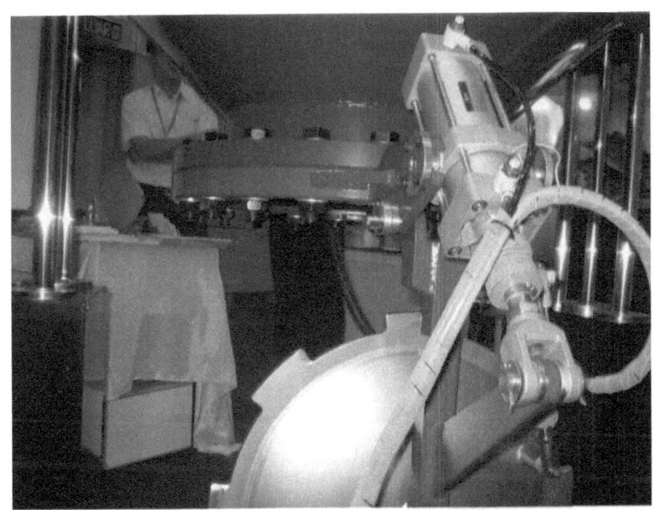

This is a pressure leaf filter displaying a flap door at the bottom of it. The flap door discharge is an excellent replacement to the existing butterfly valve as well as an improvement to the wedge lock door currently available in the market. The flap door discharge cake is specially designed to be used on a vertical pressure leaf filters. The trap door cake discharge is compact, light and easy to install. It requires no modification on the vessel and the replacement of the lipseal is quick and easy.

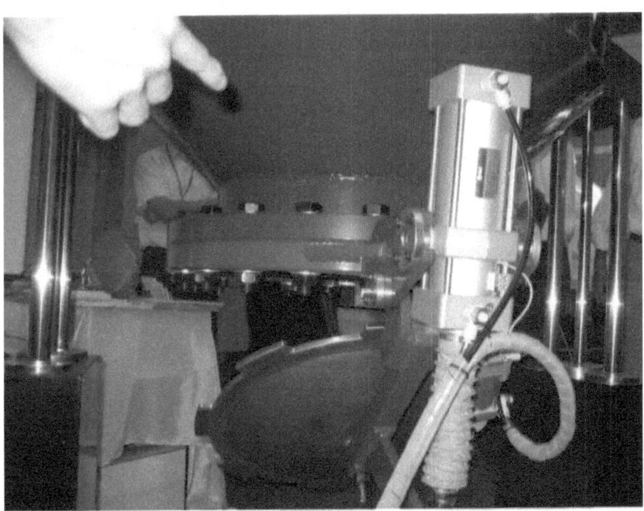

This is the same pressure leaf filter with its hydraulic flap door going to be closed. The pressure leaf filter was displayed and demonstrated by Tapis Teknik. I found this product very interesting because at my plant, we use a huge butterfly valve to discharge the cake and when the liner has problem, it is very difficult and troublesome to replace them.

I've never heard of FUNKE before. Funke is a company from German manufacturing heat exchangers. Well, I know some companies producing heat exchanger such as Smidth-Breten, Alfa Laval, Graham and Hisaka. Funke is new to me. Well, at least I learn and know that this company exist and I may get some of them later. Funke produces a wide range of plate heat exchangers.

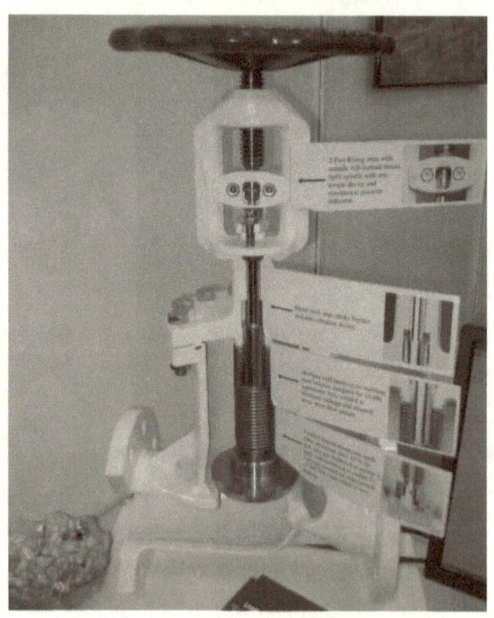

This is a 4" bellows sealed globe valve produced by W.T.A. This is the latest and one of the advanced globe valves produced by W.T.A. It can withstand high temperatures up to 400 degree celcius and has a lot more safety features.

My New Baby Plate Heat Exchanger

Since I first worked as a process engineer, I continuously learn a lot in the plant. The learning never stops. However, when it comes to heat exchanger I get very excited. I don't know why. Maybe because it is such a very important piece of equipment. Maybe because I have a lot of them in my plant. Maybe because it deserved such attention from a person like me!!! Well, there are hundreds of reasons if I have to list them down.

Sometimes, I thought I already know and understand heat exchanger. I thought my knowledge and comprehension on a heat exchanger is good enough. I'm wrong!!! Yes, maybe I know about it, but there are still more to learn and explore.

I have justified and ordered one set of plate heat exchanger (2 in series) from Schmidt Breten earlier this year to be used in my plant. I nearly ordered the wrong set of heat exchanger. I checked and checked the details and specifications of the plate heat exchanger. The purchase requirement reached the procurement department and they asked tones of questions. I understood why they asked a lot and requested me to thoroughly justify the reason of the purchase - because the plate heat exchanger is very expensive (almost similar to a brand new E-Class Mercedes Benz). I never expected the procurement executives to go line by line reading the heat exchanger specifications. After a series of discussion, I realized something was not right. I called the supplier, asked for more explanations. I discussed with my superior and found that we just ordered a wrong heat exchanger. I was scolded for ordering the wrong heat exchanger. I swiftly cancelled the order and corrected the purchase requisition.

Last week, the plate heat exchanger set arrived from German. I have patiently waited for its arrival. I have to wait for almost half a year after my request. I was so excited and happy after being informed that the heat exchanger has safely landed in the store at my work place. I immediately went to the store and looked at the new shiny plate heat exchanger. I touched it and embraced it (i looked like an idiot that time!!! Haha).

Today, the plate heat exchanger has been positioned at the installation point. It's not easy to mobilize the heat exchanger as it is very heavy. 6-8 maintenance fitters work together to erect one of the heat exchangers. I knew this is not a very big deal to other people, but to me, I'm totally responsible for the entire purchase, justification, planning, coordination, and most importantly plant process.

This new set of heat exchanger will improve my plant production back to our maximum capacity. Our old heat exchangers have suffered severe scale and fouling and resulted in slower flow rate and poor heat transfer. The overall heat transfer coefficient is very bad. The older plates need to be dismantled and replaced. Well, that's in the process. I hope everything will be silky smooth. I shall update the progress later.

"I have been impressed with the urgency of doing. Knowing is not enough; we must apply. Being willing is not enough; we must do"

Leonardo da Vinci
(Italian polymath of the Renaissance)

Some Short Updates

This occurred in mid-2008.

The pass few days were very hectic for me. Last week I gave a talk entitled *"Chemical and Process Engineering - Sharing Knowledge and Experience"* to the first year Chemical Engineering students at University Technology Malaysia (UTM). It was my first experience delivering such talk to more than 100 students. I guess it went well and I hope the students gain some valuable information from my talk.

Today, I just returned from a day trip to Kuala Lumpur. I'm still exhausted but I push myself to publish a post here. I met the design project group from University Technology MARA (UiTM) that I supervised as allied supervisor. They need some help on certain areas of the design project and I explained and assisted them. I think it's very good that UiTM introduce and include allied supervisor from industry to assist supervising the design project that the students are doing. Their report will be more realistic, accurate and they'll have better comprehension on the project.

At work, I was busy as ever. We stopped one of the plants and did a swift maintenance job to improve the plants process. I'm glad that the planning was good and jobs were well coordinated and executed. Supervisors, maintenance fitters and plant operators have done a brilliant job and the plant managed to be started smoothly after that.

In order to create a better and conducive working environment, the management initiated Super 5S which is part of the Japanese Kaizen system. 5S was already introduced earlier but it slowly faded as nobody was really monitoring it. Now, a new steering committee was formed to lead, enforce and apply 5S. I was elected as one of the steering committee and that means additional job and responsibility for me. Part of my task was to be a 5S trainer and make sure everybody is well trained. Hmm....I got to start first by keeping my desk tidy!!....

Besides that, I also received few emails from various readers (of my blog) asking questions related to chemical engineering industry, career, plant process, data, information etc. I have answered some of them and I also have not answered some of it. Don't worry; I'll answer the emails as soon as humanly possible. Thanks for your emails...

"During peace time a scientist belongs to the World, but during war time he belongs to his country"

Fritz Haber
(German chemist who received the Nobel Prize in Chemistry in 1918 for his invention of the Haber–Bosch process, a method used in industry to synthesize ammonia from nitrogen gas and hydrogen gas.)

Learning Process from Cleaning Plate Heat Exchanger

In my previous post *"Some Updates"*, I mentioned about Process Plant - Heat Recovery View. The plate heat exchanger that has been thoroughly cleaned via "cleaning in place" (CIP) could not produce the desired output temperature (but we got the flow rate). I blamed the RTD for the output temperature not correct. Although the RTD has been serviced, it still shows the same reading.

Today, we realized what was wrong with the plate heat exchanger (or cleaning program). We only conducted (CIP) for one side only (cold oil 1) i.e. the side which we assumed dirty, contained scale and fouling. The other side (hot oil 2) which we thought was clean (free from scale and fouling) was not cleaned with hot caustic (main ingredient for CIP). We found out from the newly installed vortex flow meter that the 'hot oil 2' side flow rate is low i.e. about only 54m³/hr. The 'cold oil 1' flow rate can reach up to 105 m³/hr. This shows that the 'hot oil 2' side in the plate heat exchanger is partially blocked with probably scale and fouling. The heat transfer was therefore not efficient and effective. The 'cold oil 1' temperature should reach 240°C from 105°C. However, it only managed to reach 210°C which is not sufficient for the process.

Now that we know the problem, we did CIP for the 'hot oil 2' side. After 6 hours of CIP, the flow rate became 76m³/hr which shows some sign of improvement. We'll circulate the hot caustic until the flow rate reaches at least 90m3/hr. I'm glad that we learned and experienced this. At least our understanding and comprehension, tricks and technique of handling the plate heat exchanger improved. Thanks to one of my supervisors who informed us about the slow flow rate of 'hot oil 2' which is a result of not heating up the 'cold oil 1'.

Gaskets

I attended a very interesting and informative training about gaskets. I never imagine that there are a lot to learn and explore about gaskets. After attending the training (delivered by Nipseal), my comprehension on gasket improved and I began appreciating it more. From the training, I know that there are various types and material of gaskets. Prices of gaskets vary a lot and some type of gasket is very tedious and difficult to manufacture or fabricate.

What is a gasket?

According to Wikipedia, a gasket is a mechanical seal that fills the space between two objects, generally to prevent leakage between the two objects while under compression. Other definitions/descriptions of gasket: "A flexible material used to seal components together; either air-tight or water-tight" (PartSelect.com).

Gaskets are commonly produced by cutting from sheet materials, such as gasket paper, rubber, silicone, metal, felt, fiberglass, or a plastic polymer. Gaskets save money by allowing less precise mating surfaces on machine parts to fill irregularities. Gaskets are commonly produced by cutting from sheet materials, such as gasket paper, rubber, silicone, metal, cork, felt, fiberglass, or a plastic polymer (such as polychlorotrifluoroethylene). Gaskets for specific applications may contain asbestos. It is usually desirable that the gasket be made from a material that is to some degree compressible such that it tightly fills the space it is designed for, including any slight irregularities.

Gasket is very important in a process plant. It maintains the energy, temperature and pressure in a process system. Selecting a suitable gasket is a must because it does cost money. It is also directly related to the process temperature, pressure, and type of medium (fluid or gas) and the chemical properties of the medium.

Another new knowledge that I learned is about spiral wound gasket that can withstand pressure up to 70-80 bar. It is a very interesting and carefully manufactured gasket made of stainless steel.

Let me explain about the simple/normal gasket. With reference to the illustration, the gasket is sandwiched between flanges. The property of the gasket and correct compression/tightness allows the process system to maintain its pressure and would not allow oil or gas to leak.

We install/fix the correct type of gasket before connecting the pipeline with flanges.

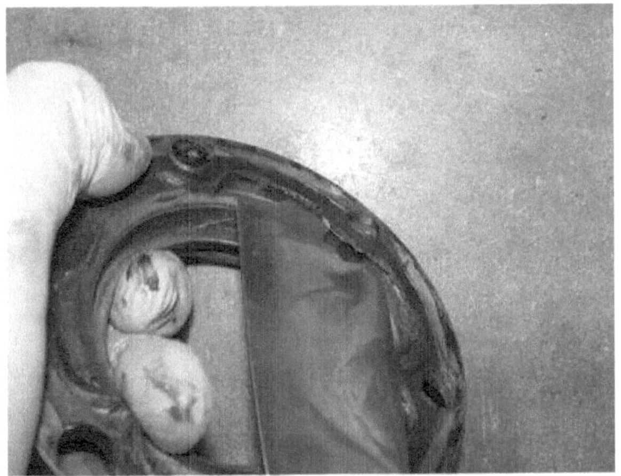

This photo shows the condition of gasket after being used for some time. It is difficult to remove the gasket by bare hands because the gasket sticks very well. We need to use suitable tools to peel and remove the gasket from the flanges. Usually, flanges like this are opened during shutdown to clean/clear pipelines or vessels. After inspection, a new fresh gasket will be used. Never use a gasket twice.

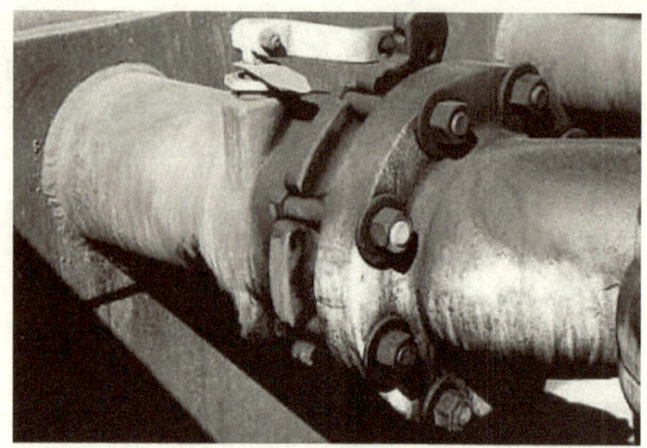

Without a gasket, a 12" butterfly valve like this will leak and spill oil or water and spray gas or steam (depending on your application).

Gasket is not only used between flanges. It is widely used everywhere in a process plant and in our kitchen (the refrigerator, toilet and sink). In a plant, gasket can be found in the heat exchanger, valves, vessel man holes etc. The gasket needs to be properly maintained to ensure no upsets in the plant.

"I've learned that mistakes can often be as good a teacher as success"

Jack Welch
(American retired business executive, author, and chemical engineer. He was chairman and CEO of General Electric between 1981 and 2001)

Disappointing New Plate Heat Exchanger

We have waited for the new heat exchanger for more than 6 months. It safely arrived last month. We managed to installed the heat exchanger and connect the pipeline. Thanks to the hard work by the maintenance fitters. The insulation work for the pipeline is in progress. That is important to avoid heat and energy lost. Everybody was eager to use the new heat exchanger. We installed pressure gauges, pressure transmitters, RTDs and temperature gauges along the new line. We fixed ball valves and bellow seal globe valves. We performed air test. Everything seemed good and promising. We finally used the heat exchanger few days ago. Unfortunately it leaked and we had to stop it. We tightened the plate heat exchanger. Ran it again, but it still leaked. We stop it and retightened it again, and ran the heat exchanger. It still leaked. It leaked at the end of the plate next to the frame. We suspected the end plate is leaking and it didn't look good. Maybe the gasket was glued unevenly to the plate surface. We were so furious because a new plate heat exchanger that just arrived from Germany is leaking. Is this a manufacturing defect? Let's see what's going to happen tomorrow.

*"If I have ever made any valuable discoveries, it has been owing
more to patient attention, than to any other talent"*

Isaac Newton
(English mathematician, astronomer, theologian, author and physicist who is widely
recognised as one of the most influential scientists of all time)

What is P&ID?

As engineers/engineering students, we always hear about P&ID. But what is actually P&ID? Well, for the benefit to those of you who are not sure on this, P&ID is piping and instrumentation diagram/drawing (P&ID). It is defined by the Institute of Instrumentation and Control as follows:

-A diagram which shows the interconnection of process equipment and the instrumentation used to control the process. In the process industry, a standard set of symbols is used to prepare drawings of processes. The instrument symbols used in these drawings are generally based on Instrumentation, Systems, and Automation Society (ISA) Standard S5. 1.
-The primary schematic drawing used for laying out a process control installation.

For processing facilities, it is a pictorial representation of key piping and instrument details, control and shutdown schemes, safety and regulatory requirements, basic start up and operational information. It also shows piping, equipment and instrumentation connections within process units in oil refineries, petrochemical and chemical plants, natural gas processing plants, power plants, water treatment and similar plants.

Process and instrumentation diagram - a family of functional one-line diagrams showing hull, mechanical and electrical (HM&E) systems like piping, and cable block diagrams. This kind of P&ID is a full diagrammatic representation of a process plant. Each piece of equipment is shown along with its connectivity to other equipment. It may be regarded as an enhanced process flow diagram which shows, in addition to the process itself, details such as control and instrumentation equipment, pump and pipe sizes etc. Each instrument / piece of equipment is shown by a symbol denoting its type (pump, sensor, valve etc.) and a unique identification number or tag for differentiation from others.

My PLC (Programmable Logic Controller) Experience

At work there are still a lot of things to be done. The latest, hottest and interesting job in my list is the PLC (programmable logic controller) upgrading in my plant and control room. For me, this is a very enriching project and an excellent learning experience. All this while, I've been monitoring the temperature, flow rate, pressure, utilities, processes, costing, equipment, instrumentations, reports etc., but now I'm going to get myself a little bit familiar with the plant control system. Previously, in university, I didn't really fancy process control; and advanced control subjects, but now, I'm beginning to show interest. However, if I'm not mistaken, the subjects don't really touch a lot on PLC stuffs.

OK, back to the present moment. Every process plant must have its own PLC to run the process or production plant. If not, the plant operators have to switch the pump, flow meter or other instruments manually and that's impossible at this era. It will be very difficult for them to control and monitor a running plant.

Let's see some definitions of PLC. According to Wikipedia: "A PLC is a digital computer used for automation of industrial processes, such as control of machinery on factory assembly lines. Unlike general-purpose computers, the PLC is designed for multiple inputs and output arrangements, extended temperature ranges, immunity to electrical noise, and resistance to vibration and impact. A PLC is an example of a real time system since output results must be produced in response to input conditions within a bounded time, otherwise unintended operation will result."

A non-technical term to describe a PLC: A PLC is the type of computer that controls machines. The PLC is used to control and troubleshoot machine. The PLC is the brain of the machine. Without it, the machine is dead.

For the past few days, I've been studying the input/output (I/O) arrangements at the control panel. These connect the PLC to sensors and actuators. PLCs read limit switches, analogue process variables (such as temperature and pressure), and the positions of complex positioning systems. On the actuator side, PLCs operate electric motors, pneumatic or hydraulic cylinders, magnetic relays or solenoids, or analogue outputs. The input/output arrangements have external I/O modules attached to a computer network that plugs into the PLC. The I/O points consist of digital input, digital output, analogue input and analogue output. I checked them and counted how many spare I/O points are available to be used for additional pressure transmitter, RTD, inverter and pump that we're going to install (which is part of some slice of plant upgrading projects). Hmmm....I guess, that's enough for some brief introduction on the PLC. I may continue about my PLC adventure in the future.

From the Chemical Engineer's wife

My husband has been begging me for quite some time already to write a post here in his blog (even though I have written 2 articles before). It's not that I don't want to, but I personally think you guys out there prefer to hear from him than from me. Out of his mountainous work load, he seems like the busiest man on earth. Despite of him trying his hardest to update, sometimes he can't catch up. So here I am, writing my two cents worth…..

I am now in my fourth semester of my Chemical Engineering PhD in a local university (Malaysia). It is not the most joyous experience for me. Since I was little, I hate studying. I have to do this one just because my faculty wouldn't let anybody off without a PhD. What a bugger. However, it is not as heavy as doing my first degree back in England as I don't have to attend classes or any exams.. what a relief. The hardest part is getting myself organized and focused.

My supervisors are both very helpful and they do not force me to do anything. They just let me be independent. I don't have to see them every week and they don't even bother if I don't see them for months…hahaha. There are goods and bads in that. The good part is that you don't have to work like a horse in meeting their target and you don't feel pressured. However, the danger is that you will be progressing very slowly. You will not be writing papers (journal, conferences), your experiments went on in a slow pace and you will be jeopardizing your PhD. Therefore, the key is discipline!! No matter what kind of supervisor you have, just discipline yourself and have your own set of targets. If your supervisor is the pushy type of person, just work with him/her and do everything he/she asked for. If your supervisor just let you be, arrange appointments to see them to discuss your targets/progress/plan. They wouldn't say no, they just want you to work on your own pace. So if you think you are not a highly disciplined person, choose a supervisor that can make you work for it.

The relationship between student and supervisor is often difficult to orchestrate. Some supervisors treat their students as colleagues and friends, others prefer to maintain a formal teacher-pupil relationship. In any case, your PhD supervisor will be an important figure in your life for at least the next three years. Your PhD will inevitably affect the rest of your career, so take some time to consider not just what and where you'd like to research, but who you'd like to work with.

Here are some tips on How to choose your PhD Supervisor.

1. If you're doing your PhD in the department where you're doing your first degree or where you are currently lecturing, just ask these questions: Does he know your name? Can you face another three years of his jokes? If you're already calling him 'Uncle Keith' it may be time to move on.

2. If you're doing your PhD in another institute:

- Approach someone whose work you know from the literature. It's important that there won't be a major clash of interests and personalities.

- Look around the department and assess your potential supervisor's standing. (if you could go)

- Talk to his other students (emails are also ok). Are they relaxed, confident and busy or do they have a glazed expression and a compulsion to look over their shoulders? Have they published single author papers? First author papers? At all?

- Communicate with him through emails. If he always answered your emails between lecture tour in Japan and a conference in Brazil, will you ever see him?

Hopefully this would be useful for those planning to do their PhD. Good Luck...

My 5-Day Break

I'm going to celebrate Hari Raya Eidulfitri Festive season a day after tomorrow. It's a big day for me and my family and the rest of the Muslims around the globe. We'll be traveling to my wife's hometown which is about 300+ km from our home. I'm taking a 2-day annual leave + 3-day public holiday. That's a total of 5 days and I'll try my best to really enjoy this break. I don't want to be interrupted by phone calls from the plant/my work place. I don't want to think about the plant. I just want to relax my mind for a while. I want some peace...

My plant will still be running this coming festive season. Luckily my senior colleague is there to take care of this very important plant. He have been guiding, mentoring and supporting me all this while and I really appreciate it. I'm learning a lot from him (and his precious experiences). On normal days, as a process engineer, I have to take care of the plant 24/7. That means to be alert and know what is going on in the plant at all time. I'm answerable on everything and anything that is happening to the plant. It's a really big and huge fast moving plant that we cannot afford any error. If not, the downtime will be very costly and I'll have a difficult time in explaining it to my superiors.

To be frank, I have to say that at all time, my mind will be focusing on the plant. While driving, going to sleep, taking bath, having dinner, outing with family, somehow, I'll think about what's happening in the plant. Is the plant OK? What if the oil quality is off spec? Is the flow rate maintained? Is the oil product sent to the correct tank? Is the vacuum pump OK? Are we getting the required process temperature? Is the steam supply enough? All these linger in my mind. Sometimes, in order to sleep peacefully, I'll call the plant and ensure everything is silky smooth before resting and going to sleep.

On a daily basis, my supervisors will call me (and my senior colleague) every early morning at 5.30 AM and report to us about the plant progress. Therefore, when we arrive at work, we are well aware of the plant overnight situation and performance. If there are problems, the shift supervisor or shift leader will call us even though it is 2.45 AM or 4.13 AM in the morning (and we are in deep sleep with sweet dreams). Well, now that I have the 5 days break - that means my mind is also free. I'll just relax and give my mind some break which it definitely deserves.

What I Get After My 5 Days Break

Yesterday, I began work after the 5 days break. It was a paradoxical emotion getting back to work while the festive season is still alive. It felt like the break was not enough. Would it be better if I can stretch my annual leaves to next Sunday. That would be a cool 9 days break for me. Well, I just have to face the fact that my annual leaves and public holidays are just 5 days. I have a number of responsibilities at work that requires me to get back to work early. One of the plants was stopped due to lack of raw material/crude oil and we were forced to take the opportunity to do some maintenance job.

Preparing a 5 days daily production reports after the break was not a speedy one as usual. It took me longer hours to prepare the report due to some changes in production planning. Having an annual management review meeting after lunch yesterday did not help me to reduce my job load. In fact, I received additional tasks and reports to be prepared within 48 hours. Up till now, I haven't completed the report yet. Well, I have to rush for it after this.

I also have to prepare for tomorrow's maintenance coordination meeting. Wow..., there are really a lot of meeting here and there. With the plant stoppage and maintenance going on which I have to monitor from time to time, I tried my best to manage and coordinate as best as I could. It's really a challenge for me to handle a lot of the projects, tasks and other jobs all at one time. Luckily, I'm still OK. We have good executives and supervisors to assists us executing the jobs. So far the plant operators and fitters are doing fine although manpower are still limited and lacking due to some of them are still on their annual leaves (remember, it's the festive season...).

I appreciate all of these as a good training experience for me. I hope other junior engineers/going to be engineers anywhere can be confident and strong enough to face any situations when they're really tested up to the maximum. They must be mentally tough and able to lead and keep track with the expectation from their superior. Anybody want to share experiences...?

Disappointing New Plate Heat Exchanger Is Now OK...

This post is the continuation of the "Disappointing New Plate Heat Exchanger" post.

After installing and using the new plate heat exchanger (PHE), we could not get the flow rate and temperature. This was really a problem because we were expecting the new PHE to perform excellently.

The PHE seems to be leaking although we have tightened it from 540mm to 529mm. The minimum length we can go according to the design is 506mm. But that is applicable after running the PHE after certain duration of time, not tightening it while it's still new.

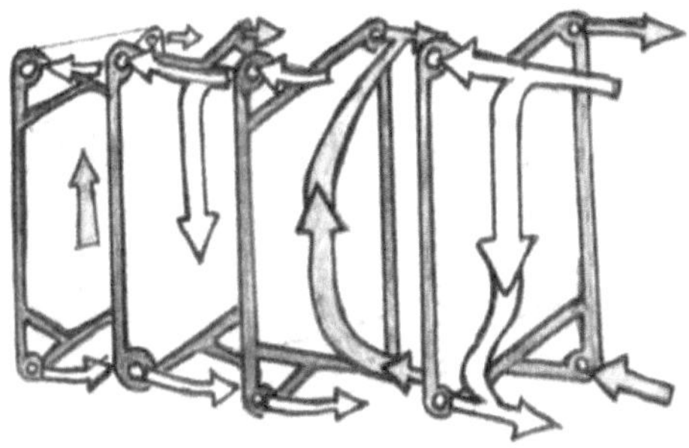

I called the PHE supplier and asked what's wrong with the PHE? What should we do? Should we tighten it some more? The leaking is only at the end plate. Should we dismantle and check the end plate? Maybe the gasket is not evenly glued on the end plate? Maybe the gasket is damaged or distorted!

The technical engineer came the following day and we performed an air test to show him the leaking points. He noticed the leaking point and agreed to dismantle the PHE to check what is wrong.

After dismantling the PHE, we found out that one plate was not positioned correctly. That was very surprising and it was not supposed to happen for a branded and reliable heat exchanger from Germany. Soon we found out that the heat exchanger was actually fixed locally by the supplier's principal and not by the manufacturer in Germany.

The technical manager then guided our maintenance fitters to fix back the plate heat exchanger. Carefully they checked the plate's arrangement. Diversion plates, flow pates and end plates must be in correct order. We don't want to repeat the same mistake. If not, we are just going to waste our time and energy.

After completed fixing the PHE, we conducted air test and hold the pressure at 2 bars. We used soap liquid to check for any possible leakages. The pressure maintained for nearly one hour and that was a good sign that the plate heat exchanger is not leaking.

We then gradually used the PHE and finally we got the desired temperature and flow rate. I was so happy and glad that the new PHE was working perfectly as we planned. Now, we can focus on other improvement and maintenance jobs such as cleaning in place (CIP), insulation, spare pipeline, and others. We're taking it one by one.

Moral of the story:

1. Don't expect a product or service to be perfect. Hope it to be perfect! If not, trouble shoot as soon as possible.

2. Be extra careful while arranging and dealing with plate heat exchanger. Once your arrangement is wrong, the entire effort is a waste.

Production Plant Problem

Normally, there'll be less or no problem when a production plant is running smoothly. The supervisors and plant operators will be happy taking care of a normal running plant. The executives and engineers will monitor and optimize the processing parameters and ensure all utilities consumption such as water, electricity, steam, air, chemicals, natural gas, LFO, diesel etc are kept at the lower side. Once in a while there'll be some problems such as leaking mechanical seal in a centrifugal pump or the level transmitter signal is not showing the right reading. Well, those are just some minor maintenance problem and could be easily entertained by the fitters and technician.

However, the big test is when the plant is stopping (or starting). A plant which is running smoothly will be interrupted and stopped (vice versa). The utilities consumption will be higher than normal. The temperature, pressure, flow rate will be disrupted. The plant operators will adjust certain processing parameters and also a few valves in order to stop the plant safely. The supervisor must properly and carefully coordinate the plant stoppage (or starting). At this point, the experience of the supervisor and operators plays a very significant role. Operating the stoppage (or starting) of the plant just by referring to the working instruction or manual will not be sufficient. For me, I'll be confident if the stopping (and starting) of the plant is led by an experienced staff/supervisor. It's even better if a senior executive or engineer could be around to monitor and oversee any problems or possible danger.

There were cases where serious accident/disaster occurred during starting up (or commissioning) of a plant. In a case 2 years ago in a plant next to my work place, the distillation column was caught by a blazing fire leaving 3 staffs crying helplessly for assistance on top of the highest roof above the column. We thought those unlucky staff would lose their lives either by being burnt to death or from injuries after jumping down the 80 meter column to the ground (if they could not stand the heat). Luckily the fire brigade finally came and used their ladder crane system to fetch the trapped staff from the almost melting structure. There must be something that had gone terribly wrong during that particular plant start-up. That's why we need to be extra careful and have ample manpower to assist on the plant stoppage or starting up.

From The Chemical Engineer's Wife Part II

In my previous post, one of the readers put a comment requesting for some tips on choosing PhD specialization. Well, I am obviously not the most suitable person to talk about this for I myself is still in the process of pursuing the degree. However, I will try my best to incorporate my own experience on how I choose my specialization/topic here.

My Experience

I was forced to do my PhD. As soon as I finished my Master's degree, I started teaching in a Chemical Engineering Department at a local university. I enjoyed my teaching years where I can get close to the undergraduate students and try to make them understand the subjects. Unfortunately, a Master's degree is not adequate for me to supervise a Master's degree or/and PhD student. A Master's degree is also not the faculty's goal for all the faculty members. A PhD is inevitable in order to continue teaching and working as a lecturer there.

Therefore, with a heavy heart, I filled the form to enroll in the PhD program in my own faculty. It was not hard for me to choose my supervisor because I choose a person that I have worked with before for my master's dissertation. I know his personality and he knows mine. We clicked. However, at that particular time, my supervisor has run out of research grant money and has already completed all his existing research. Therefore, I was assigned to choose my own topic. I was lost for a while…but not for long. I reflected on what I want to do in the future and where will I be positioned in the faculty. I am supposed to be specializing in 'separation method' in chemical engineering because I have been told to join the Separation Group of the faculty. To be more specific, I chose 'crystallization' as my specialized area because nobody in the faculty has mastered it.

During my Master dissertation, I researched on Palm Oil Crystallization in Producing Edible Oil. Therefore, I decided to continue in the same area but in different medium. I searched the internet on crystallization and two of the areas caught my attention, *metal glass crystallization* and *freeze concentration*. After reading existing journal papers on both topics, I decided to do *freeze concentration*. The reason: it's cheaper, easier to understand and interesting. To conclude, the choosing process was not difficult for me because I know where I'll be positioned in the future in my career.

For those who are just embarking in pursuing their PhD, here are some tips on how to choose a PhD topic.

Some Tips on Choosing a PhD Topic

1. First of all, do it with considerable care. Never lose sight of the fact that the PhD thesis should be the crowning achievement of your graduate education and will influence the direction of your career for many years to come.

2. It is of great advantage if the supervisors are willing to offer a choice or two or three dissertation topics. He/she will have a better overview of the field, knows the sources, and knows if the dissertation is doable within the allotted time frame. In effect, you will receive a crucial implicit promise that you will be closely guided along the way.

3. If you have good reason to be confident in doing research on the topic of your own choice if close guidance feels too restrictive to you, then proceed, but at least be forewarned that you can easily lead yourself on a wild goose chase.

4. It is imperative that both you and your supervisor be interested in your thesis topic. It is important that your mentor be interested in it because otherwise he/she might be much less motivated to help you.

5. Make sure that you do not start a dissertation on an unfamiliar topic. You should prepare some plans, even if tentative ones well in advance and have a good overview of the topic before you commence active research.

6. Once you have chosen your thesis topic in collaboration with your supervisor, you should seek his/her active guidance to the utmost degree possible.

7. You will need to learn who the important scholars are in the field. Ask your supervisor who is working in your area, check their respective home pages on the internet, and look for their working papers.

8. Whether you take a topic selected by your supervisor or develop your own, you have to be excited about the topic. I think it is more likely that this will happen if the topic is developed by you yourself, and coming from questions that you really want to pursue.

"This most beautiful system of the sun, planets and comets, could only proceed from the counsel and dominion of an intelligent and powerful Being"

Isaac Newton
(English mathematician, astronomer, theologian, author and physicist
who is widely recognized as one of the most influential scientists
of all time)

Dipping Tape and Measuring Oil Tonnage

In oil industry or oil plant, there are a lot of oil storage tanks available. It is vital for the oil company to know their daily oil stock simply because they are in the oil business. Therefore, they need to measure the oil volume and further work out its tonnage. They need to know the exact volume of oil and to do that there are various methods. One of the traditional methods is by using a dipping tape.

What is a dipping tape? Sorry, I could not find a suitable and proper definition for it. My own definition would be a tape made of from stainless steel (can be other material as well) and has some sort of calibration along it (up to 15m, or 20m, it depends on your tank height) and it's used to measure the ullage in a storage tank (filled with oil/liquid). Does anybody have a suitable definition of dipping tape?

I guess the next question from you will be: What is ullage?

The amount which a tank or vessel lacks of being full. It is the empty space present when a shipping container is not full. It can also be defined as the space in a tank not occupied by its contents. Used as a measure of storage space still available. I hope the ullage definition would give you some idea on the concept.

Dip weight - It is attached to the end of the tape

For each storage tank, there'll be some sort of calibration table. From it, we can know what the volume of oil at certain height is. The tank height is also recorded in the calibration table. So, in order to know the volume of oil, we subtract the tank height from the ullage (which we get from the dipping tape measurement), and we shall get the oil level height. From there, we can get the volume of oil and work out the tonnage (with the temperature reading available).

"Learning never exhausts the mind"

Leonardo da Vinci
(Italian polymath of the Renaissance)

300, Design Project 2, 5S, NPSH, Pump - Rambling!!!

I did not realize that the previous post about "Dipping Tape and Measuring Oil Tonnage" was my post number 300. That makes this current post that you're reading - number 301 in this Chemical Engineering World blog. Wow, it has been about 1 year and 3 months since I began blogging about my working experiences. I never expected thought or dreamed this blog can be like what it is today. I never imagined having more than 250 subscribers (at this moment and growing) for a technical chemical engineering blog. Thanks for your support.

This coming days will be a very hectic one for me. Tones of jobs and projects are waiting for me at work. We are going to have a minor 9 hours plant maintenance job. That job must be carefully coordinated in order to reach its objective - which is to improve the vacuum system in our plant process. I also have my plant monthly reports to complete and that is not easy as it needs me to really focus and concentrate.

At the same time, the final year design project 2 file number 2 just arrived last Friday from a local university for me to read, digest, check, evaluate and give marks. I'm not sure when I can finish reading the design project report, but I was ordered to complete and email the marks by the end of this week... Well, I got to rush for that as well. I almost forget that I also need to conduct 5S training for my plant operators. Maybe I'll arrange that on Tuesday afternoon.

Yesterday, on Saturday, I attended a very brilliant 3 hours training regarding pump. The training was conducted by one of our very experienced and knowledgeable superior. From the training I discovered various types of pumps with various applications. We also did some real practice on how to calculate the Net Positive Suction Head (NPSH), based on our plant condition and situation. This was more advanced from the basic NPSH that we learned in university. Furthermore, at university, we only know the theory, but this time, we experience and appreciate the NPSH knowledge... of course after we faced all the pump problems in the plant. We did also learn on other stuffs on why the pump trips, cavitate, air lock etc. and its relationship with the pump curve.

Lately, I also received few emails asking me some favours on various issues and also asked for some advice/assistance. I'm trying my best to entertain those requests, but occasionally, lately, I'm too busy. I hope you can understand my situation. I'll try my very best to get back to some of you. I'm also struggling and fighting with time to keep this blog updated with fresh feed and content for you to read. That's why I really appreciate anybody who is willing to provide one or two post (or more... which is good) so that we can have this blog updated more frequently. If you have anything to share, it can be from your own experience or some new stuff that you learn or know, you're welcome to email it to me. If it suites this blog, why should I not post it in this blog for you?!!

I also have some post which I did mention to elaborate more. I'll continue them soon. Just bear with me for a while. Some of you have personally asked the continuation of those posts. Sorry for the delay mate...

"Science can amuse and fascinate us all, but it is
engineering that changes the world"

Isaac Asimov
(American writer and professor of biochemistry at Boston University.
Known for his works of science fiction and popular science.)

Me and My White Safety Helmet

When I was a student, I wonder how it would feel to become a chemical engineer. How does it feel to wear the safety helmet that will make us look like a real cool macho engineer?

In the lab at the university, we just put on the white lab coat and not the safety helmet. After my first degree, I continued with my master's degree and still, I have not yet put on the safety helmet. I wonder when will I own and wear the safety helmet.

After completing my master's degree, I joined a local oil and gas servicing company and directly travelled to the site on my first day at work. I was supplied not only with the safety helmet that I wanted to put on all this while, but also jacket, safety boot, coverall, cotton glove, leather glove, goggle and 3M half face mask. On top of that, I have to manage a group of people who were about 5-10 years older than me to blend some specialty chemicals. Under the hot shiny sun, I have to wear the safety helmet. That time, I wished I don't have to wear the safety helmet with the goggle attached on top of it (The goggle need to be applied while blending or pouring the chemicals). I don't feel like the safety helmet is protecting me from anything. After all, there's nothing going to fall onto my head at the side. It was really heavy and I felt like my centre of gravity is at my head!!!

Despite all of that, I felt sort of proud to display my safety helmet on the rear dashboard of my car and let everybody see it. At home my son will take my helmet and act as an engineer, just like his father. I don't know whether he wants to be an engineer too. It's totally up to him.

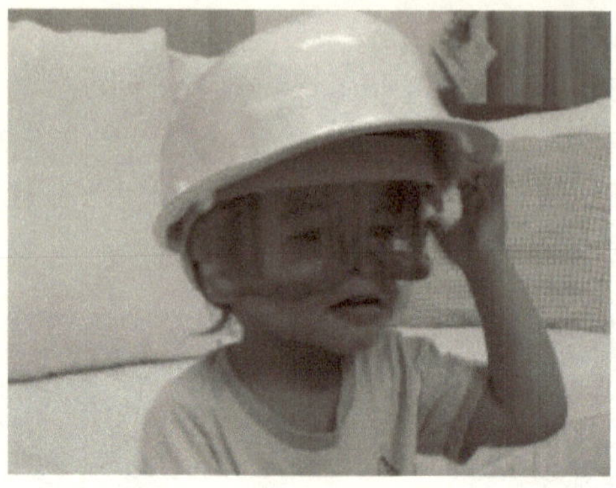

Few years after servicing the oil and gas industry, I got a new job as a process engineer in a refinery plant (in the oils and fats industry), I still put on the safety helmet. Now, it is different. I don' have to put any accessories on the safety helmet. It is lighter. I have to wear the safety helmet which is white in colour as soon as I enter the factory. White safety helmet differentiate executives from supervisors, technicians, operators, and others which put on a yellow safety helmet. In different places, safety helmet colour coding is applied. Some may have blue colour representing a safety committee or safety officer etc. In brief, the colour of safety helmet indicates certain specific roles of the person wearing it.

Wearing safety helmet in the plant is very important. The plant is really big and there are many equipment. I lost track of the number of occasions where I knock my head on something. Sometimes, my head hit a metal bar, a globe valve, a lowered roof and others which I could not recall. Luckily I have my safety helmet to protect my head and brain! I could not imagine the state of injury I'm going to face if not protected by the safety helmet.

Previously, I wrongly used the safety helmet. How did I misuse it? I sat on it!!! That is totally a wrong usage of the safety helmet. As a result, my safety helmet has a lot of ugly scratches. The white safety helmet on the picture above belongs to me. If you have super focus eyes, you can zoom in and notice some scratches and uneven surface on the helmet edge. Don't be like me. Appreciate your safety helmet. Don't sit on your safety helmet. Respect it. It will protect you, your head and brain!

"Man is essentially ignorant, and becomes learned through acquiring knowledge."

Ibn Khaldun
(Fourteenth-century forerunner of the modern disciplines of
historiography, sociology, economics, and demography)

Production Plant Problem # 2 - Inconsistent Pump Flow – Part 1

We can expect problems to occur while starting and stopping a plant as these are the most critical period in operating a plant. One of most popular problem is having an inconsistent pump flow. The following is a sharing based on my experiences in those plant start-stop moments.

Problem # 2 – Inconsistent pump flow

The pumps below a buffer tank could not deliver the required flow rate that they were supposed to. The flow rate was extremely slow and the discharge pressure was not consistent at all. The pressure went crazy up and down from 0.5 to 5 bars. We tried to adjust and play around with the pump in order to get the desired flow rate.

Slow flow rate means lower throughput, hence lower production. In actual, if we keep on running with the low flow rate, production will be 50% less. We could not afford that to prolong. We have heavy shipment ahead and we need to immediately rectify the pump problem.

There were all together 3 pumps in a row of similar motor kW and pumping capability. Initially we thought all three pumps were having some problem (or the same problem). We asked the maintenance fitters to check, service and replace the mechanical seals. We checked the pump pressure and found it was OK because the discharge line pressure can go up to 4 – 6 bar. There's no way the pump is having a problem.

Then we thought it was Net Positive Suction Head (NPSH) problem. We increased the level of oil inside the buffer tank so that the NPSH will increase. That didn't work either. The flow rate was still low and the discharge pressures were inconsistent.

By this time, we suspected the pumps were experiencing serious air-lock. Air-lock is a situation where the pump could not pump efficiently and effectively due to some disturbance from air turbulence inside the pump suction line. The pump discharge line pressure inconsistency might be due to the air-lock problem. The question now is where the air is coming from? We decided to thoroughly check the pump suction line to search for any hole/leak (which air may easily enter because the tank is operating under vacuum). We removed the insulations covering the suction line, checked and inspected it, but still we could not see or detect any leak (because the tank and pipe was under vacuum, which means oil would not pour/come out from the leak due to vacuum holding it). Time was running fast but we have not yet settled the problem. We began feeling the pressure coming from our superior/management.

After almost one day running low, we made a more drastic and radical move. We hold the plant and braked vacuum – stopped production. We waited for a while. At first, we noticed nothing. The pump suction pipeline looked fine. After 15 minutes, we were caught by surprised on what we saw! Droplets of oil came out from the tank insulation. At first, it was little, but then there were more and more oil pouring out. This confirmed that there is a leak somewhere underneath the tank insulation. We cautiously removed a small portion of insulation covering the tank where the oil came out (because the oil and the surface was extremely hot (260°C)).

After successfully removing the insulation, we saw oil pouring out from the welding joint between the tank and the pipeline. We looked at each other, wandering how we are going to resolve the already identified problem. We cannot simply weld the leaking point because traces of vacuum were still inside the tank and system. If the maintenance fitter welds the leak point, we fear spark may enter the tank (and system) and trigger fire inside the tank, pipeline and other vessels. At the same time, it is impossible to weld the leak point because oil was coming out from the tank. Should we drain the balance oil in the buffer tank? If yes, that would cost a lot of time, estimated half a day. Just imagine the downtime we already faced and add up another half a day for repair work and starting up the plant again. We have to think of a better and faster way to weld the leaking point.

Finally we came out with a very risky idea but can solve the problem faster. We have to maintain a slight vacuum inside the buffer tank just to hold the oil from pouring out from the leaking hold.

To be continued in a few days' time...

"What we usually consider as impossible are simply engineering problems... there's no law of physics preventing them."

Michio Kaku
(American theoretical physicist, futurist, and popularizer of science)

Production Plant Problem # 2 - Inconsistent Pump Flow - Part 2

This post is the continuity from Production Plant Problem # 2 - Inconsistent Pump Flow and Production Plant Problem.

Finally we came out with a very risky idea but can solve the problem faster. We have to maintain a slight vacuum inside the buffer tank just to hold the oil from pouring out of the leaking hole. We must attain the right balance i.e. to avoid oil from coming out of the vessel so that welding can be done; and to avoid spark from entering the vessel to avoid fire. We gathered enough manpower to execute and monitor the job.

The welding job runs smooth. Due to lack of time, we could not conduct air or steam test to really ensure that the welded portion is properly sealed. With that in mind, the fitters must properly weld the leaking section. They welded a few layers to ensure there'll be no leaking at all.

While the welding job was in progress on the ground floor, another big vessel on the first floor was externally caught on fire. Smoke rushed out from underneath the insulation and we can see fire coming out of it as well. We swiftly grabbed the fire hose and spray water towards the fire.

It was the insulation wool that was actually burning. The big vessel has some hidden leak and when the plant breaks vacuum, oil came out of it. Combination of oil, high temperature and air triggered the fire as the flash point was reached. That was not the first time. It had happened several times during plant startup or stoppage. Luckily, we managed to put off the fire. However, we continued spraying water to cool down the hot insulation wool. We also poured few pails of low concentrated caustic to encapsulate all leaked oil (hydrocarbon molecules) in the insulation wool.

As soon as the situation on the first floor was under control, the welding job on the ground floor was also completed. Everything looked fine and the plant is ready to start. The vacuum system was slowly established and crude oil is pumped into the plant. After I was really satisfied with the overall situation, I left the plant, relieved that we have identified the prime pumping problem that haunted us all these while. It was already dark and I was very exhausted. Phewwwhh....

Chemical Engineering Related Photo Sharing

I'm going to share some interesting photos with you. I'll first let you see, observe and analyse what those photos are. Just imagine and think what are the equipments or situations... It's better if you figure out what the photos show before I reveal the answers. If you are a student, you will find this a good learning curve, I hope so. If you are already working in a processing or production plant, you may already know about these photos.

(a) What is happening?

(b) Another view on what is happening in photo "a".

(c) Another clue on what is happening in photo "a".

211

(d) What is this?

(e) Another close up on photo "d".

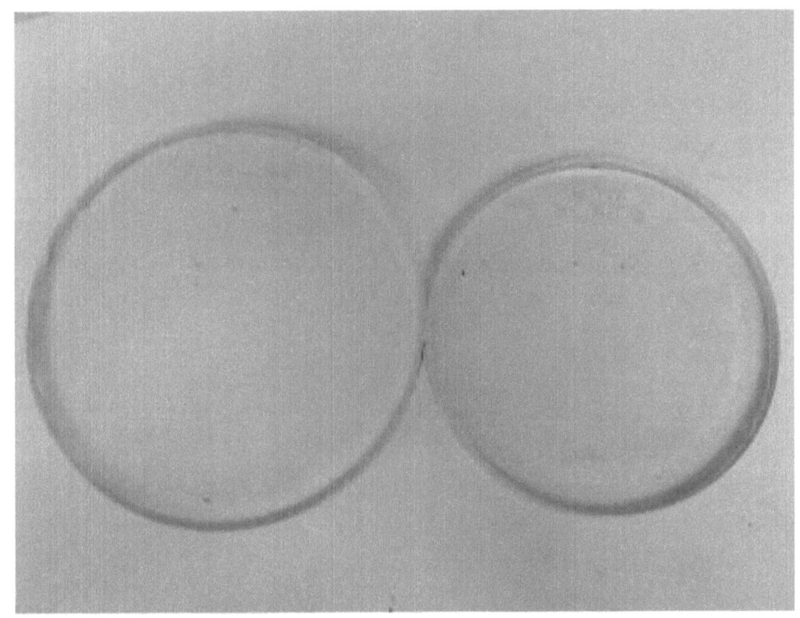

(f) What is this?

(g) Another clue on what is photo "f"

Answers:

Photo a: A plate heat exchanger is leaking.
Photo b: You can see traces of fluid leaking/pouring out from the bottom of the plate heat exchanger.
Photo c: An example of plate heat exchangers' plate where fluid may come out when the incoming hot and cold fluid are not balance. Or, if the gaskets are no longer functioning due to damage, torn, not perfectly attached or various other reasons.

Photo d: This is the ball which originated from a 4" 3 piece body ball valve. It is made of stainless steel SS316.
Photo e: This ball was taken out from a damaged 4" 3 piece body ball valve. This type of ball valve is imperative in processing plant. It is more reliable than a gate valve and butterfly valve but is more expensive.

Photo f: These are sight glass. It is used to view and observe physically the flowing fluid in a pipeline or stream.
Photo g: This photo is an example on where the sight glass is installed. By having a sight glass, it is easier to monitor and observe the condition and physical appearance of a fluid.

"Excite yourself with short term goals that will enable the accomplishment of your ultimate long term goals."

Zaki Yamani Zakaria
(Founder of Chemical Engineering World Blog &
Chemical Engineering World FB Page)

Plate Heat Exchanger Is My Favourite Equipment

For this one month period, I'm going to focus my post more towards my favourite equipment in a processing / production plant - plate heat exchanger. I am very passionate and possess high interest in this equipment. A plate heat exchanger does not use electricity and that's why sometimes it is also called economizer. The heat energy is exchanged from the hot fluid to the cold fluid in very well and carefully arranged plates (between plate heat exchanger frames). How much kilocalaries or kW energy transferred really depends on the fluid inlet properties, flow rate and the plate heat exchanger design.

For a start, I'll list out all of my previous plate heat exchanger related entries for your references (available in this book).

Heat Exchangers: Introduction and Basic Concepts
Chemical Engineering Related Photo Sharing
Learning Process from Cleaning Plate Heat Exchanger
Preparation Work for New Plate Heat Exchanger
My New Baby Plate Heat Exchanger
Disappointing New Plate Heat Exchanger
Disappointing New Plate Heat Exchanger Is Now OK...
How Do You Analyze Your Heat Exchanger Performance?
Fixing Plate Heat Exchanger Gasket
High-Temperature Heat-Transfer-Fluid Systems

Sometimes, I feel like switching my job and work for a heat exchanger company and make myself a plate heat exchanger specialist. But that position is very limited.

Interrupted Planning

Sometimes or rather often, it is difficult to plan our work. What I mean is the work depends on various situations and decision makings. We planned our annual leaves and vacations but suddenly something important came in and we have to cancel the entire plan. Have you been facing that?

Cancelling your plans after planning them for weeks or months must be really frustrating. But, we have to accept the fact. We are working for somebody. We are an employer. We are an engineer. We have a responsibility and we must deliver what we are supposed to do.

So, what can be very important that suddenly hinder you from being temporary away from your routine work? There are a lot. It depends on what field you're in. It can be a problem faced by your customer and you got to fly urgently to their place and solve whatever problem they faced. I've experienced such thing before. A problem appeared while my colleague was handling a small routine project in Kerteh. There was some trouble blending our chemicals and it affected our customers' production. My boss rang me and asked me to drive to Kerteh that night to clear the mess. As a committed employee, what can I say.....? 30 minutes after the telephone conversation, I began my journey from Johor Bahru to Kerteh (6 hours non-stop drive), leaving my family behind over the weekend. That was from my previous job as a chemical/project engineer.

As for my present job, I really look forward to enjoy a short one day break during the coming Labour Day. However, to my surprise I'm not going to be able to enjoy it. We have a problem. The main problem is simply lack of raw material. At the same time, the annual machinery inspection is due. Hence, both reasons combined and resulted to a week-long annual shutdown beginning this Sunday. There goes my weekend and my public holiday (This is what I call "area beyond control"). All the planning and decision was made in a space of 5 hours. And within those 5 hours, I've to rush to arrange and plan everything.

Tentative shutdown planning, schedule, manpower arrangement, contractors, JKKP (my colleague is handling this) and others need to be settled before immediately. Why? Because tomorrow is going to be a weekend and the annual plant shutdown is going to take place commencing this Sunday. Friday afternoon is the only time to settle all the planning and coordination. Luckily, I've pre-planned the shutdown and I can swiftly prepare all the necessary stuffs for the shutdown.

Well, that is some slice of my feeling, emotion and story that I can share. I'm not saying that I'm not happy working as an engineer. I like the job and I'm very proud to be an engineer. However, this is what we call responsibility. As an engineer, we have to perform whenever and wherever we are required to. At some point, there are a lot of sacrifices that need to be done. It's just a matter of how well we can adapt and cope with our job. Are you tough enough to be an engineer? Only you can answer it...

"Knowledge without action is wastefulness and action
without knowledge is foolishness"

Al-Ghazali
(Prominent and influential philosophers, theologians,
jurists, and mystics of Sunni Islam)

My Chemical Engineering Career Overview

After graduating, I have been working in the research and development area, oil and gas industry, and oil and fats sector. That's altogether 3 different fields. To be honest, I'm glad I went through all those fields. It made me taste a little bit of something from those areas.

I never planned it to be like that. I believed God has arranged all of these for me. God wants me to be more diversified, more all-rounder and know more technical details from various industries. Now that I'm going to turn 32, I think it's already time for me to decide on which area to dwell. I need to establish myself in an area where I can grow together with it and become the expert in it.

My wife is a chemical engineer as well and after completing her degree, she straight away joined a local university as an academic staff. She continued her Master degree in Chemical Engineering after a year becoming a tutor. After getting her master, she became a lecturer, teaching Unit Operation and Mass Transport subject. After a while she was on study leave because she was doing her Ph.D.

As you probably already know, my journey differs from her. If you have been following my blog from the beginning, I have shared my R & D experiences, my oil and gas journey - going onshore and offshore - traveling in and outside the country, my oil and fats adventure - taking care of a 3000 MT/D capacity physical refining plant etc.

Well, there are pros and cons on my part and that is exactly what I want to share. Check on the following points:

Pro

1. I tasted and experienced a lot. That means I know generally more about those industries.

2. I have been working in several offshore platforms which I believe not many chemical engineers have done that. Traveling offshore is very exciting and I always look forward for it. The allowances are great and if you are the kind of person who wants real physical challenge and want to make big bucks, by all means, try and apply for a job in an oil or service company.

3. I experienced working as a service person and I also worked as an end product person. Those are totally different. You might want to know about this before you work. In the 2nd point above, I mentioned about applying for a job in an oil or service company. That will reflect your working lifestyle.

If you work in a service company, you'll be more flexible and probably do a lot of traveling. You are doing services to your customers. Hence, there'll be a lot of traveling here and there. I made a really crazy trip few years back from Johor Bahru to Jakarta to Johor Bahru in one single day. I was in Skudai in the morning, headed to Batam, took a flight to Jakarta, and met up with our company principle. Then I flew back to Batam, took a ferry to Johor Bahru and at 7 pm I was already back in Skudai eating burger with my wife. It was a really tiring journey. The travelling time was 12 hours, via land, see and air.

If you work in an oil company or end product company, you'll be normally stationed just there. You have to take care of a plant, or project or maintenance etc. That place is going to be your world. You'll be there from morning until your working time ended. Most of the time, you have to work additional hours because there are simply tonnes of work to do.

Cons

1. I could not focus to one area of work as I switched courses 3 times in a space of 9 years. However within that small space of time, I tried and worked hard to become a good employee each and every time.

2. Jumping here and there delayed me to climb the corporate leader. Some of my friends have become senior engineer or manager. However, I'm still a normal process engineer. That's fine. It's OK. I'm still having a good job and decent pay.
What I wish...

If I have the opportunity to choose which industrial career based on my present knowledge and information, I would like to work in either 2 areas:

i. A Chemical Engineer in a servicing company in the Oil and Gas industry. Example of the companies - Halliburton, Schlumberger, Clear Water, Weatherford etc.

ii. A Technical Engineer for a service/product company selling heat exchanger. To be exact, I would like to work for Alfa Laval.

However, I would also like to be in the academic line and involve in research and development. But, I'm not sure yet about that. Let's just see how my career is going to be...

What about you? Are you working your dream job?

"Simplicity is the ultimate sophistication"

Leonardo da Vinci
(Italian polymath of the Renaissance)

Are Our Local Universities Producing Poor Quality Engineers

I stumbled upon this article from the Star Online Newspaper. It covers the frustration of an employer towards the fresh chemical engineering graduates released from local universities. This is a very interesting fact that we should seriously address. Our young engineers must be equipped with soft skills and able to deliver when they are expected to do so. Check out the article and tell me what you think about it.

Engineers of Poor Quality

I AM a manager in a chemical manufacturing firm in Malaysia. We often have vacancies for mechanical and chemical engineers, and occasionally electrical engineers. We do take in fresh graduates to train and develop for the future of our company.

In recent years, I have noticed a marked reduction in the quality of the engineering graduates. I would like to suggest that our local universities work with professional bodies such as Board of Engineers Malaysia (BEM) and Institution of Engineers Malaysia (IEM) to address the weaknesses.

Some of the courses should be tailored to suit industrial requirements. BEM and IEM would be in the right position to work with the many universities we have here. Alternatively, they could come up with modules to be included in the engineering curriculum at our local universities.

With the advent of computers and simulation packages, another new problem is that fresh engineers seem at a loss to conduct design calculations from basic principles. They are over reliant on such computer packages.

When they start work, they are at a loss to do design work because some companies may not have such computer packages. Hence, even basic engineering calculations to determine the optimum pipe sizing and pump selection are beyond them.

These are basic engineering calculations, and without the necessary skills, we are left with design works that are sub-optimal, resulting in high operating costs for the users.

Alternatively, everyone would be running to consultants to get even the most basic of engineering work done for them. In many of the plants I have been to, there is much that could be done to improve efficiency by just going back to good basic engineering practice. And in some cases, it's just using good common sense.

I think there is a need to teach and emphasize on such basics. We should ensure that our young engineers are provided with good foundation knowledge for the future of our country.

After all, it is upon solid foundations that skyscrapers are built.

In this aspect, I must take my hat off to University Technology Petronas (UTP), which has formed an Industry Advisory Panel (IAP), and invites professionals from the industry to review their curriculum and suggest areas for improvement. UTP is serious about this and has implemented many of the suggestions introduced by its IAP.

UTP also has an adjunct lectures series where professionals are called in to give lectures to the undergraduates. I think these are good initiatives that other universities would do well to emulate.

SHYAM LAKSHMANAN,
Lahat, Perak.

Adopted from the StarOnline Thursday April 17, 2008

My two cents:

From my observations, a lot of newly graduated engineers are lacking of the confidence they are supposed to possess. Engineers come and go from various work places. Most of them resigned because they cannot perform, cannot cope or cannot stand working. They are not mentally tough. Two years plus ago, a group of engineers including me was scolded badly by our big boss. The next day, one of us resigned. He simply said that he could not work in that kind of environment. He cannot work under immense pressure from the superiors.

However, some of the newly graduated engineers are well performing. I admit that. I am so impressed with a few colleagues of mine who performed excellently as an engineer. They adapt and communicate very well. They effectively coordinate and manage project and work. As for myself, last year, I supervised a practical student. I told him that we are going to communicate in English all the time. He was not allowed to converse in other language or our mother tongue language. The reason was for him to practice conversing and communicating in English fluently. We did talk and discussed in English. I think he was a good student and perform all the exercises and assignments given to him by me. I bet he'll become a good reliable engineer someday.

Being a Malay working in a Chinese company seems quite challenging. However, I have no fear and problem with that. Although I can't speak or understand when they converse in Mandarin, I ensure myself conversing in English with all of them. On a daily basis, I will attend a meeting with a group of Chinese executives, engineers, chemist and managers and I'm the only Malay in the group. When they talked to me in Malay language, I replied in English. I don't give a damn. Some suppliers also acknowledge this. They were impressed of me because I still converse in English and forced them to speak the same language with me.

Well, those discussed above are mostly related to the soft skills. When it comes to the technical skills, as engineers we must show what we are capable to do. We must be able to do all those basic calculations and engineering work. Most of the time, we will have to learn more simply because we have not been exposed to those new areas or knowledge while at the university. We have to learn and show that we can pick up whatever projects or tasks assigned to us. My previous boss will always encourage me to meet up with suppliers because they are the expert in their field. We can call and meet up with them, discuss and increase our technical knowledge. Frankly, I learned a lot from my suppliers. I think it is not too harsh if I say, we utilized and used them for our sake.

Bottom line is, new engineers or present engineers, we need to work out on our soft and technical skills. Do not disappoint our boss although sometimes you might hate him. After all, we must show that we are worth what we are being paid. At the end of the day, if we perform, we'll get a good increment and bonus!!!

Power Surge from Local Power Supplier

There are simply a lot of problems in a processing or a production plant. I did mention a broad or general idea on the problems before this.

It will be very frustrating if a problem arises from an uncontrollable source. For example, today (somewhere in 2008), there was a power surge which came from the local power supply. It happened so fast and we felt the power interruption in a split second. Immediately, I looked through the CCTV and checkout what was happening from inside my office. The plant operators were running here and there checking the valves, pumps and other equipment. Three units of boilers which are supposed to supply steam to the plants shut off. The boiler man worked very hard to start up the boiler step by step. My colleague, a charge man, swiftly checked the LV (Low Voltage) room to check on any problem or tripping on our site. Fortunately, everything was in perfect order. It was the 1 second power surge that affected us and definitely resulted in a terrible downtime to our production.

How to avoid such downtime?

For our side, at the moment, we cannot avoid this. It must be controlled from the local power supplier, in our case, Tenaga Nasional Berhad (TNB). They must ensure their local power station is performing well and delivering sufficient power to surrounding factories and plants. Well, that's one kind of downtime that is very costly and unavoidable. Problem beyond our control.

Some Problems We Find in a Processing & Production Plant

As a process engineer or production executive, we shall always hope and wish that the plant operates smoothly. A processing or production plant will always have a problem whether we like it or not. What will the problem be? Are they taught in the university? Can you get them in the text books? I bet you would not get that valuable information anywhere there?

So, how can you get them? You'll actually get them when you work and experience those problems yourself. Another way of getting that precious information is by asking and learning it from experienced executives and engineers.

I've been working for almost three years in my current work place and I observed all these problems. I think it might be useful if I share all those plant problems with you guys. There are a lot of problems and I posted them in my blog from time to time. The problems vary and come from various angles and areas such as (without any particular arrangement):

1. Utilities problems:
Power, natural gas, steam, water, chemicals, LFO, diesel, processing aid, air etc. Processing cost can increase the overall production cost.

2. Maintenance & Equipment problems:
Pump, piping, instrumentation, pressure transmitter, level transmitter, temperature indicator, NPSH, cavitation, control valve, steam trap, leakage, insulation.

3. Human resource problems:
Disciplinary, tardiness, absence, failure to obey instruction, negligence, psychology.

4. Report & Documentation problems:
Daily report, monthly report, quarterly report, yearly report, ISO & GMP related documents.

5. Communication problems:

Miscommunication, instruction, network, PC, server, bad relationship with up line, down line and colleagues.

6. Quality problems:
Laboratory, lab checking, quality control, testing error, solution/chemical contamination.

7. Supplier problems:
Raw material - product - goods out of spec, cheating.

8. Supporting equipment problems:
Deterioration of cooling tower performance, heat exchanger performance.

9. Planning problems:
Administration interruption, supply demand, market, margin.

10. Control system problems:
Supervisor Control and Data Acquisition (SCADA), Human machine interface (HMI) and PC, IT, network, softwares.

11. Stock, raw material, storage tank problems:
Network, false information, over flow, insufficient storage tank, contamination.

12. Stress, Pressure & Health Problems :
Meeting deadline, lack of time, inadequate knowledge, 24 hours alert and standby, meetings, reports, various crisis, inability to manage pressure from top management.

That would be some very general and surface introduction on what problems we can expect from running a plant.

Things That I Learned From a Simple Bolt and Nut

During the previous shut down, I allow myself to become more hands on. Why? You can read on to know why in case you haven't read them. There's a lot to learn from a simple bolt and nut. There are various problems and difficulties that we can possibly encounter from a pair of bolt and nut. Followings are few things that I have learned from that small item (which are more inclined to maintenance).

1. If you over tightened those bolt and nuts, you might have some difficulties to open it later in future. If things get worse, the bolt and nuts can break. If it breaks, it will be difficult to open/dismantle the flange. In certain cases, you need to cut those bolt or nuts using oxy-cutting. However, we can only use oxy-cutting for bolt and nut made from mild steel or galvanized iron (GI). Stainless steel bolt and nut normally don't have any problem to be loosened because they do not corrode/rust.
Morale: Do not over tighten the bolt and nut. Just tighten it according to your normal human power.

2. Over paint. Sometimes you paint the bolt and nut for certain reason. One popular reason is to avoid rust. One problem that our team faced during the previous shut down is to unscrew the nut. Oh my God, it was very tough because the paint covering the bolt and nut was so thick. We applied paint remover but still the stubborn paint would not completely be removed. As a result, we took about 3 hours just to remove an 8" flange with 8 bolt and nuts. This affected our time and productivity.
Morale: do not over apply the paint.

3. Unsuitable length or size of bolt and nut. There were cases where we lost some bolts and nuts due to poor housekeeping. When we want to replace them, we need to get suitable bolt and nuts. A shorter bolt will be required for a simple flange. However, a longer bolt will be required to sandwich a check valve between flanges.
Morale: Use the correct size and material of bolts and nuts.

4. Grease the bolt and nut. After removing the bolt and nut from a flange, it's better to clean them if there is any debris or rust. Then apply grease to avoid rust and to ease screwing and unscrewing.

Morale: Get a cup of grease and ask your operator/fitter/worker to apply the grease.

5. Ensure there is sufficient stock of bolt and nuts. During a shut down or turn around, we normally dismantle flanges and manhole to clean vessel, pipeline, distillation column, deodorizers or others. Sometimes, a small number of those bolt and nuts broke and they need to be replaced. We must have ample stock of the correct bolts and nuts so that we can smoothly connect the flanges or close manholes.

Morale: Ensure your store has sufficient stock of the required bolts and nuts.

6. Tightening the bolt and nut. When tightening the bolts and nuts, we need to criss-cross the arrangement. Normally a flange will have 8 holes for the bolts. Commence with 12 o'clock, followed by 6 o'clock. Then tighten 3 o'clock followed 9 o'clock and so on. This will ensure even compression towards the gaskets when we tightened the nuts.

Morale: Do it patiently and correctly. If not, the system may leak and we'll get in trouble (that is if we don't do a steam or air test before starting the plant).

7. Prepare the correct size of spanner. It's a waste of time if we didn't use the correct size of spanner to unscrew a nut from a bolt. If you want to unscrew a nut of size 24, use spanner size 24. If you want to use an adjustable spanner, use the one that is closer to the size range. Do not use an adjustable spanner which is too big. If you use an oversize adjustable spanner, you will get tired faster or you might trigger an accident.

Morale: Ensure you have the correct size of spanner when dealing with bolt and nut.

One Vital Plant Shutdown Lesson

We just completed our 7 days plant shutdown and it left me with a mixed feeling. In my earlier writing, "Interrupted Planning", I highlighted about how the plant shutdown suddenly took place and it had interrupted not only my plans but also others. Before this, I really look forward for it. Luckily I haven't planned anything special or big during that public holiday. Whenever the plant shutdown was instructed to be conducted, I already knew that I'm going to be in trouble. I had already approved annual leaves for 3 of my downline staff (2 supervisors and 1 shift leaders). They have taken leave for 4 straight days from Thursday (the Labour day) till Sunday. It waas difficult for me to ask them to cancel their annual leaves because they have planned the leaves 3 or 4 weeks ago, applied the leaves and I have approved it.

The problem also occurs in Maintenance Department, who plays significant role in our shutdown also, they had some manpower shortage with similar reason. So, we had to arrange our limited manpower accordingly to do all the routine and special jobs for the annual shutdown. To make things worse, another 2 of my manpower (1 my senior supervisor and 1 operator) were sick and unfit to work. The senior supervisor had a kidney stone while my plant operator had a high blood pressure up to 170. As a result, the number of manpower I had was slashed down again.

It was a really serious challenge for me. I was having a shutdown but I didn't have my key people to assist me. As a result, I had to make myself an engineer cum supervisor. I walked around the plant, checked and assisted all works. I managed our contract worker manpower as well. I cannot imagine what's going to happen to all those work if I wasn't there to coordinate and monitor the jobs. I personally did some of the shutdown work such as filling up the high pressure boiler with deionized water, working on those flanges to dismantle or tightened the nut and bolts etc. I became a physical hands on person.

It was a very expensive lesson for me. Next time, I cannot simply approve annual leaves for my down line staffs. They can take leave, but whenever there are very important jobs like a shutdown, they need to sacrifice their leaves.

IEM Southern Branch Technical Visit to Claytan Group

Last month, we (me and my wife) joined the Institute Engineer Malaysia (IEM) Southern Branch for a technical visit to Claytan Group. It was my first technical visit with IEM and I began enjoying my association with the institution. This technical visit is worth 4 CPD hours (BEM approved). Altogether there were about 30 professional engineers joining the visit.

Here is a little bit of information about Claytan Group (adopted from their website):

"Founded in 1920, Claytan Group manufactures high quality ceramic products ranging from vitrified clay pipes, sanitary ware, art ware, tableware to hotel ware. Combining nearly a century worth of experience and cutting-edge ceramics manufacturing technology from Japan and Europe, Claytan emerged as one of the most experienced Malaysian ceramics manufacturers with the most diversified ceramic product range in the country.

Claytan Group products are designed with the customer in mind. They strive to find innovative ways to improve the quality of our ceramic products to produce products that rank high on customer satisfaction. The quality of our products is recognized by many international Quality Assurance (QA) bodies such as SIRIM, PSB, AS and CSA. They carry out third party audits twice a year to ensure that our quality management system and products meet their stringent requirements."

These are some of the gorgeous tableware shown in their display room. These are products from Claytan Group. Most of the beautiful ones are exported. Some of them are sold locally.

More elegant and modern looking tableware exhibited. Do you like them?

Technical presentation from Claytan Group representative. All IEM members were tentatively listening to what he was presenting. We were the youngest engineer participating in the technical visit.

Some of the clay pipes manufactured by Claytan Group. We visited the plant and learn their processes which began from various clays being crushed in a crusher. We witnessed their efficient conveying system and large mixers. Then there is a very huge kiln to bake and harden the clay pipes.

There are also various tests carried out on the clay pipes to ensure the quality is up to the required standard. One of the tests is the hydro test which is to test whether the clay pipe can withstand its designated pressure or not.

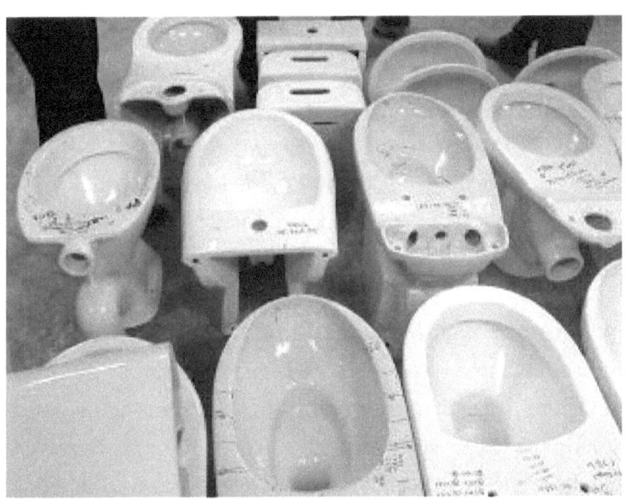

These are different batches and designs of toilet bowls being tested in their R and D department. Honestly, before this, I never imagined a toilet bowl being studied as extensive as this. There are also a lot of test being perform to ensure the toilet design works well and can remove the "sludge" in it. I also learned that there are 2 types of toilet bowl, one which can operate with 4.5 litres while the other one with 6 litres of water. It was also interesting to learn about the vacuum type flushing toilet which is also used in a plane.

I was really impressed with all the hard work made by the research and development department to ensure they produce a high quality toilet bowl that can work efficiently and effectively. It was also my first time touching the holes of the water outlet from the bowl. I never touched it before. I never thought of touching them. I touched it to inspect the size and diameter of the holes where the water poured out during flushing. Very interesting.

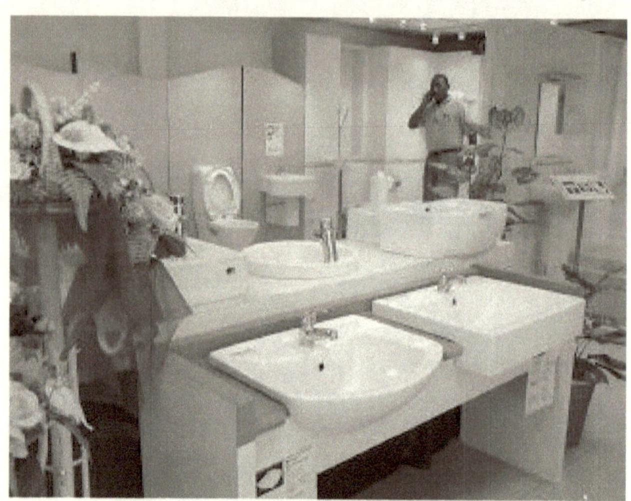

These are part of the sanitary ware exhibited at the display room. Claytan Group is a major player of the country's sanitary market.

Conclusion

It was a very informative and beneficial visit. Besides listening to the technical presentation, checking out their sanitary research and development department, visiting their clay pipe plant, they also served us good warm food for lunch and gave us a very nice colourful mug (as souvenier). We also visited their wholesale tableware and saw a lot of their collection. I sincerely thank them for their hospitality to us.

For those who are not yet associated with IEM or any professional engineering association in your country, I urge you guys to do so. It is very good for your future. You can also mix and mingle with other fellow professional engineers and hence expand your network. Cool stuff.

Technical Visit to NEWater Facilities Singapore

My second visit with IEM was to NEWater Facilities in Singapore (pronounce it as New Water). The NEWater process recycles our treated used water into ultra-clean, high-grade reclaimed water, cushioning our water supply against dry weather and moving Singapore towards water sustainability. NEWater is managed by Public Utility board (PUB) Singapore. It was a really interesting technical visit and I wanted to learn about their process as an idea because I am handling the project to purify effluent water to be used for our cooling towers and washing facilities. The visit gave me some ideas on how I should approach to build our own micro or ultra-filtration system.

The half day technical visit took place on the 24th of May 2008 on a nice beautiful Saturday. We gathered in front of the IEM building and got into a specially chartered bus to bring us all (about 30+ engineers from various disciplines) to Singapore.

As soon as we arrived, we were welcomed by their officers and brought for a technical tour. We were so impressed with their modern and high tech facilities. They have also spent substantial amount of money to educate Singaporeans and others to learn how they process the water. The education and learning centre was very impressive and anybody who have visited their centre would agree with me. We were given a bottle of NEWater drinking water as a souvenir - some of us immediately tasted them while others kept it. Well, no worries....the water is extremely clean and safe - Thanks to the Reverse Osmosis treatment system.

"We are continually faced by great opportunities brilliantly disguised as insoluble problems."

Lee Iacocca
(American Engineer and automobile executive)

These are sample of the first batch of NEWater bottles being distributed. There are various types of bottles and label designs for NEWater. NEWater drinking water is not for sale. They are distributed on certain occasions. Visitors to NEWater facilities will be given a Newater bottle as a memoir.

This is part of NEWater reservoir. Nice...Cool...

Presentation by one of the NEWater tour guide. They have done great job in explaining to us about the history, news, progress, technology, process, future etc.

A simulator to educate the visitors on how much water we used a day based on our lifestyle. It reminds us to appreciate clean water and do not waste them unnecessarily.

I never thought they made us see the control room. This means anybody; everybody can see the control room and its operators working in it. Cool.

Some of the piping of the system is made from stainless steel pipeline. The plant is very clean and well maintained. Well done.

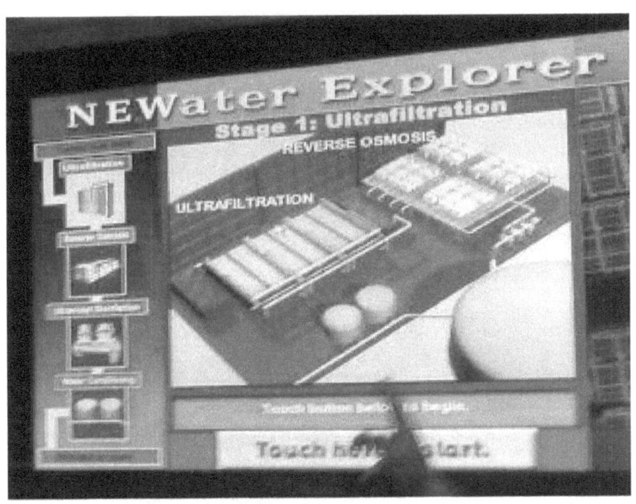

LCD TVs like this are prepared in the NEWater facilities to provide visual for better understanding to visitors.

These are the example of membranes for micro filtration of the water. Micro filtration takes place before proceeding to reverse osmosis stage.

These are the casing for membrane filters exhibited. Water from outside the membrane will be forced to enter the membrane pores, leaving out bigger particles to provide cleaner water.

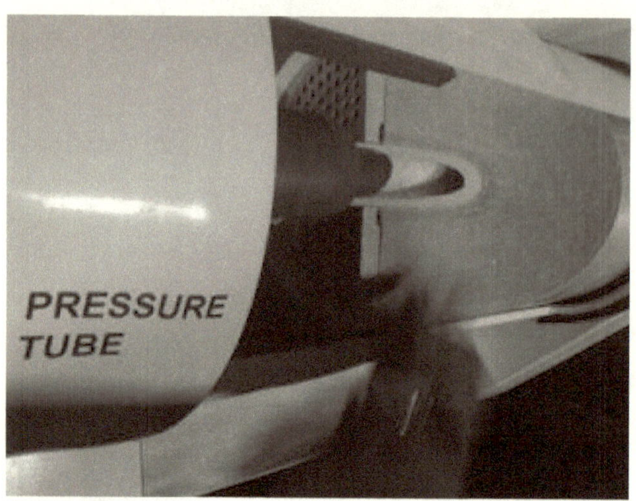

This is the membrane for reverse osmosis. Water from the micro filtration stage will enter the tube and forced to enter the membrane by exerting pressure onto it. The pure water will then enter the core of the membrane and channel to subsequent stage - UltraViolet process.

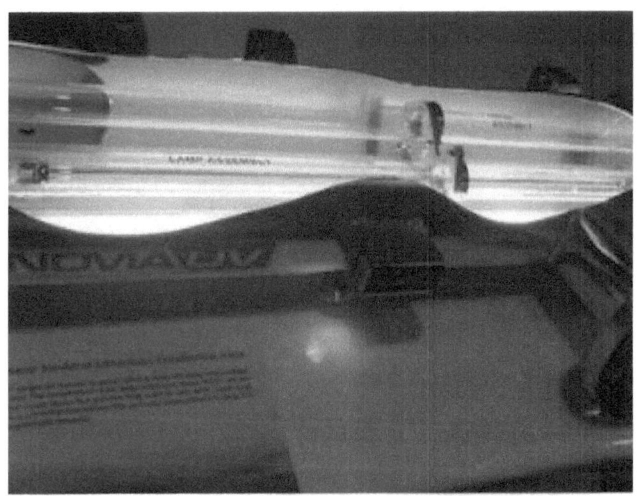

This is the ultraviolet step which is to terminate any nano sized bacteria that may possibly pass through the previous micro filtration and reverse osmosis system. This ensures us that the water produced from NEWater is extremely clean and safe for consumption. One tube of ultra violet canister will have 4 tubes of UV lights.

Oh yea... That's me signing off from the NEWater facilities in Bedok, Singapore.

For more information on the NEWater, please check the NEWater website. If you are interested to pay them a visit, they also welcome you to do so. You can call them or book online at their website to arrange for the visit.

Engineer Training Engineer - What is the Benefit?

In a space of 3 months, I taught and trained 2 newly graduated chemical engineers about my physical refining plant. As a person in charge of the plant, I thought I already know everything about my plant. However, having those 2 new engineers really tested my real understanding of my own plant.

The two of them were fresh graduates and they haven't seen or experienced being in a plant before. So, they were very curious about the plant, its process, equipment's, vessels, Piping and Instrumentation Diagram (PID) drawing and a lot more. I explained to them the process involved in the plant step by step with reference to the plant PID drawing. As I went through the initial stage i.e. from the storage tank, then to pump, strainer, heat exchanger and so on... both of them (in different occasions) tentatively listen and made notes. Then we went through the Niagara filters, filters bags, filter cartridges.... and so on. Then I explained about all the heat exchangers in the deodorizing section. Then to the packed column, another series of heat exchangers, filters and to product storage tank (I'm not trying to explain my process here, that's just a brief idea on how I explained the steps and processes to them).

The PID contains all the piping, valves, actuators, pump, RTD, pressure transmitters, level sensors, flow meters, heat exchangers, vessels, strainers, and filters in the plant. I need to really look at it and trace the lines in order to show the flow. That's good. At least it refreshes me when I'm explaining to them.

In several occasions, they asked about this and that. Some of the questions were easy to answer while some others were surprisingly difficult to answer because I forgot, or I don't know about it. That means there are still stuff about my plant that I haven't covered yet (that I don't know). They made me realized about it and it made me seek for answers. Hence, after investigating about it and getting answers, I informed those new engineers and indirectly my comprehension of my own plant became better. Looking at the scenario in a different angle, I'm actually revising and learning about my own plant when I teach or explain to others. Isn't that cool? Both parties benefited. They learned new process and engineering knowledge while I established my understanding on my plant.

Morale of the story: Don't keep the technical knowledge to yourself. Share them. The more you share, the more you'll get. The more you share, the better you are. The more you share, the stronger you become.

"I am, and ever will be, a white-socks, pocket-protector, nerdy engineer, born under the second law of thermodynamics, steeped in steam tables, in love with free-body diagrams, transformed by Laplace and propelled by compressible flow"

Neil Armstrong
(American astronaut and aeronautical engineer who
was the first person to walk on the Moon)

Submersible Pumps

I was fortunate to have the opportunities to operate several types of pump such as submersible pump, drum pump and centrifugal pump. Here, I would like to highlight about submersible pump that I have used in the oil and gas as well as oil and fat arena. Both submersible pump and drum pump are mobile type, which we can carry and operate on site with the availability of electrical power. We normally call submersible pump as sub-pump. It's easier to pronounce.

A submersible pump is one that has a sealed motor, fitted in a pump body. The total assembly is immersed in the fluid that needs to be pumped. The main advantage of this type of pump is that it can offer a considerable amount of lifting power, as it does not rely on external air pressure. A submersible pump has a system of mechanical seals that is used to prevent the fluid from being pumped from entering the motor, resulting in a short circuit. The submersible pump can either be attached to a pipe or a flexible hose.

Some of the types of submersible pumps are bladder pumps, bilge and ballast pumps, borehole pumps, booster pumps, and centrifugal pumps. Other examples are condensate pumps, dewatering pumps, fountain pumps, grinder pumps, micro pumps, sampling pumps, trash pumps, utility pumps, and well pumps. Some submersible pumps are manufactured for particular applications. These pumps are water submersible pumps, sewage submersible pumps, 12-volt submersible pumps, sand submersible pumps, irrigation submersible pumps, and solar submersible pumps.

Submersible pumps are found in many appliances. Single stage pumps are utilized for drainage, sewage pumping, common industrial pumping, and slurry pumping. Multiple stage submersible pumps are normally used for water abstraction. These pumps can also be found in oil wells. Moreover, submersible pumps can be positioned directly in a pond and requires comparatively little installation. These pumps are also relatively silent. The four main specifications that should be considered while choosing a submersible pump are maximum expulsion flow, maximum discharge pressure, horsepower, and discharge size.

Being Energy Conscious

As a process engineer working in a private company previously, I'm very sensitive to energy saving and cost reduction. I manage a big physical refining plant and I'm answerable to the production cost. Management will ask me why the natural gas cost is high. Why the electricity cost is high? Why the water and steam cost is high? Why the processing aid cost is high? Why the salary is so high compared to last 3 months? Why the maintenance cost is double that of last month? Why the average production capacity is 10% lesser compared to last year's? Those are the type of questions that have always been bombarded to me and the others who manage a plant.

Hence, I would tell my supervisors and operators to be more alert on the energy issue. I provided them with the understanding of high production cost and the direct relation with energy saving. High production cost will make the company loose money. That will lead to lower profit, lower salary adjustment and lower annual bonus or incentives. Those are among the explanations that we can provide to them.

We can also explain about the green energy and also the global warming crisis. However, sometimes, being an operator, their academic background may not reach the level of understanding we hope they have. Hence, by explaining global warming and green energy would not get them excited. They will become blur and just grin while you are explaining to them about those global issues. Hence, by associating themselves with the company performances, salary and bonus, it can make them think and react accordingly. Always the motivation is the dollar sign.

As for me, the idea of saving energy and utilities are well permanently stored in my mental box. The practice is not just at work but also applied at home. At home, we are the one who pay the bills and we tend to be more disciplined. We realize the importance of utility savings. An example is using an energy saving bulb (5 kW) for our lamps instead of a normal bulb (40kW) and also boiling water using gas instead of electricity (I do that). My electricity bill is always below RM50 per month. You need to analyse your bills as well. Check on how many kW you are using. Check on how many m^3 of water you are consuming.

Oh yes, by choosing a well-ventilated house, I don't have to use the air conditioner. I have lived in a house where the ventilation is very poor and I hated living in that house. It was so hot and we need to use the air conditioner a lot. Using a simple 3 bladed fan is sufficient and breezing. After 3/4 months, we clean the fan blades to remove the dirt sticking on it. This will improve the efficiency of the fan and the fan ampere won't be too high - We save power...

But, what about those who work in a place where they are not really concern about energy?

"Engineers ... are not superhuman. They make mistakes in their assumptions, in their calculations, in their conclusions. That they make mistakes is forgivable; that they catch them is imperative. Thus it is the essence of modern engineering not only to be able to check one's own work but also to have one's work checked and to be able to check the work of others."

Henry Petroski
(American engineer specializing in failure analysis,
a prolific author, a professor both of civil
engineering and history at Duke University)

If I Can Travel Back to the Past...

It has been almost 20 years since I received my Chemical Engineering Bachelor Degree from University of Bradford, United Kingdom. Sometimes I thought myself on what should I have done differently if I can travel back to the past. What I will do if I can repeat the entire life as a chemical engineering student? What changes will I make on my lifestyle as a chemical engineering student? What will be the improvements that I can make upon graduating as an engineer?

After thinking about it, I came out with the following points:

1) If I can travel back to the past, I will put more effort on my studies. I want to consistently do well in all the tests and examinations. I want to excel with flying colours and get first class honours. I don't want to do last minute study. It is not going to be easy but I must be strong and disciplined in order to do well.
Moral: Excel in your academics.

2) If I can travel back to the past, I will make sure to join a very big multinational oil company - Esso (currently known as ExxonMobil after it merged with Mobil) as an intern or trainee engineer. I was so disappointed to miss the opportunity provided by Esso to join them on summer 1998 for 3 months. The offer letter was sent to my home in Malaysia, but I was studying in UK. Unfortunately, my father informed about the offer after a few days. By the time the offer letter reached me in UK, it was too late for me. I missed a very good opportunity to expose myself to the real engineering world.
Moral: Make sure you take advantage of any industrial training available. You'll be surprised on how the real world is.

3) If I can travel back to the past, I want to take additional classes on process control which will allow me to have a degree called: "Chemical Engineering with Process Control". I would like to have added value in my chemical engineering degree. I should have listened to myself, my heart and not following my friends.
Moral: Make sure you know what your goals are and stick to your plans. Do not simply follow others.

4) If I can travel back to the past, I want to put more effort in my final year design project and get better marks for it. My project back then was *"Alumina from Bauxite"* and we have to design and build a plant to process bauxite until it becomes Alumina. It would be better if we can visit a similar plant in Scotland to have better comprehension on all the processes and equipment. We should have insisted our supervisor, Prof N. Harnby to arrange the visit for us. In my team, I have to design a rotary drum filter. Unfortunately, I wasn't able to imagine how it looks like although I have read and seen the diagram inside various related text book. That time the internet was just kicking off and there are lack of information on the process and the equipment. I know I cannot blame the internet. I need to work harder to get more information for my assignment. If I have better network with my seniors or practicing engineers, it would be a different story. I can ask or check with them. But, I don't know anyone from the industry...

<u>Moral</u>: We need to work extra hard in order to fully understand the subject. This is a chemical engineering course and it is not an easy course. You'll be proud being a chemical engineer...believe me... because I'm proud being one.

5) If I can travel back to the past, after graduating, I will straight away register myself with Board of Engineers, Malaysia (BEM) and Institution of Engineers, Malaysia (IEM). I will straight away get myself a mentor and follow all the training programs. By doing that, I can be a professional engineer at 26 or 27 years old. By associating myself with fellow professional engineers, I can extend my networks and also exchange/share technical knowledge. When I was doing my degree, I have become a junior member of Institution of Chemical Engineers, UK (IChemE) and upon graduating; my membership was upgraded to "Associate Member". However, I wish I had followed all the training programs in order to become a chartered engineer. The constraint that time was the high cost of the renewal membership fee. Luckily, now we have IChemE Malaysia branch and the fees are charged in Malaysian Ringgit instead of Pound Sterling.

<u>Moral</u>: Make sure you register with related professional bodies, association or institution and follow the training to be professional engineers. Nobody stresses the importance to me before, hence I ignored it. Now that I'm already following the training program and realized the importance of becoming a professional engineer, I keep advising young graduates to start early.

Well, those are the few points which I can share with all of you. I hope you can learn something from it. I wish somebody had come to me and share something if not everything about what chemical engineering is all about. Now, I'm sharing a slice of my career experience for the benefit of young and future chemical engineer.

"Willingness to change is a strength, even if it means plunging part of the company into total confusion for a while"

Jack Welch
(American retired business executive, author, and chemical engineer. He was chairman and CEO of General Electric between 1981 and 2001)

I Resigned From My Engineering Job

I wanted to write about it a few weeks ago. Unfortunately, I was too tight up with various tasks and issues that need to be settled first. I have resigned from my job as a process engineer. My last day working as an engineer was 30th June 2008. The next day I was officially working as an educator (I like to use educator more than lecturer because it sounds better). I would like to take this opportunity to share my experiences on the few last days as a process engineer and my first few days as an educator.

Previous job chapter....

After being offered as a lecturer I tendered in my resignation letter to my superior and Human Resource Department. It was really a mixed feeling that time. Is this really true? I really wanted to be a lecturer and I've been trying hard for 5 years. I could not believe it when I received the offer letter in my house mailbox. I thank God for providing me a chance to fulfil my wish...

What happened in my previous work place...? Oh...my plant operators gave me a Parker Pen. I haven't used it yet because surprisingly there is no ink in the pen!!! Maybe they just bought the pen, wrapped it up and gave it to me. BUT the most important thing is the thought. I missed them and I believed they missed me too... :)

On my last day, my boss arranged for a surprise lunch. I did not expect that. We enjoyed the lunch but I did not eat very much because I was sitting exactly in front of my boss! So, you know the feeling. For me, I found it difficult to eat peacefully in front of him.

After lunch, with some assistance from my wife, she helped me ordered 8 pieces of 12" pizza online. It was a lot cheaper ordering online compared to calling them. You should check it out if you still don't know about it. On top of those pizzas, I ordered "pisang goreng (fried banana) and cendawan goreng (fried mushroom)" delivery. All those pizzas, fried banana and fried mushroom were a courtesy from me… a person who is going to leave the company… I treated my friends, operators, fitters, technicians, store people and others who have been very cooperative to me during my stay at the company.

At 5.30 pm on that Monday (my last day at job), I shook my hand with my office colleagues. But when I shook my hand with one of my senior colleagues, Mr. Gopal, which has been close to me and mentored me a lot… I felt very extremely sad. Suddenly I felt so sad that I could not help it. Without realizing, tears dropped while I was driving home leaving the company I served for 3 years and 22 days. It wasn't a long time but honestly I felt like working at that company for 10 years. I'm finally free from that type of "refinery pressure"…

Before this, I have to be 100% alert all the time on everything that is happening at my plant. I have to know and ensure smooth operation of the plant 24 hours a day and 7 days a week. Whenever there is a problem in my plant, I need to ensure that it is settled as soon as possible. I must answer my superior and sometimes to the headquarters for any problem or downtime occurring in the plant.

Please don't be mistaken. I'm not saying that I don't like my job as a process engineer. I liked it a lot, very much indeed. I've learned various aspects of technical knowledge, project management, management skills, manpower management, social, political issues throughout my position and my stay there and I've been sharing some of them here. However, I think I can contribute more when I'm in the academic line. I can educate the young going-to-be engineers, I can do research work, I can become a consultant, I can do publications, attend conferences and the list simply goes on. I'll be more flexible and I know it is going to be a better career for me.

Being an engineer for almost 6 years has given me the real exposure on how to be a good profound engineer which is indispensable for the company. Hopefully, I can educate the students better based on my experiences in the oil and gas field and also the refinery life.

My New Job

I have been in my new job for almost a fortnight. It was quite a solid packed 13 days as I was assigned many tasks. So far the job is not really relaxing but it is still better than my previous job as a process engineer which has more intense pressure. I realized that my new job as a lecturer/educator requires me to be more disciplined compared to my previous job. We have the time flexibility and we must know how to manage it wisely. We cannot misuse the flexibility given to us. I have my own individual office room, and it is meant for people like me to work more effectively and efficiently. I have my own large working station, sofa, huge rack etc. I just requested for a 2 drawer metal rack to lock my important stuffs, just in case. My wife's office is just 20 steps away from mine… ;). Oh yea…we have the magnificent wifi facilities and that means I'm connected to the net all the time…

However, besides all that, the real challenges for me are preparing for the classes and also deciding on the research I'm going to do for my Ph.D. I need to do some extensive survey on what research I should conduct for my Ph.D this coming semester. This is going to be my future because my career will be based on what research niche area I'm indulged in. I need to be an expert in certain area so that I can easily develop that area without competing with anyone locally. So far, I have brainstormed, searched and listed a dozen of research title. Most of them are directly related to renewable energy and reaction / catalysis engineering.

Special note: Although I'm no longer working as a practicing engineer in a plant, I'm still an engineer serving in the academic and research arena. I also associate myself with the industry and other engineering institution society. It is my goal and objective to share my knowledge and experiences to engineering students, young engineers and other practicing engineers so that they can benefit positively.

This is a newspaper cutting from a local paper "Utusan Malaysia" 5th July 2001. This is when I was doing my Master in Chemical Engineering - "Single step conversion of methane to gasoline using zeolite catalysts".

Concluding Remarks

You have read my selected ramblings in my blog that I wrote since 2006. It comprises of my mixed experiences in three distinctive industries. Throughout the journey after I completed my chemical engineering degree in 1999, I appreciate all the experiences attained. I cherished them. It was a daily exploration of knowledge. It was a stimulating voyage. Every day I got excited learning various new things. I learned it from my colleagues, technician, engineers, executives, suppliers, managers, bosses, mentors and others; and of course a lot of self-study (online and offline).

The early years of any engineer's career is very important. This is the starting point. This is the point where you established your career, begin to learn something in much detail and specialize in it. Where your career begins will manage to keep you interested, will possibly be your life career. Hence, you need to choose your early career wisely, if you have the choice. Consult more experienced engineer for tips and advice. Some young engineers are unfortunate because there is lack of opportunities in their respective country. My advice is never ever give-up. Keep on pursuing your dream to be an engineer.

As I highlighted in this book earlier, my experiences commenced when I was involved in the (i) research and development niche, then followed by (ii) the oil and gas arena, then (iii) oil and fats sector, and (iv) finally academia world (which I do not cover in this book). All of this is a process to make me a better engineer, a matured engineer and finally a professional engineer. This process must be gone through by all engineers in order to be a successful one that contributes to the technological society and world.

Later, I discovered that my journey as chemical engineer does not end in the academia world. There are more, more that I discovered, more than I expected from the continuous years of service. In 2016 I embarked in a managerial role as an administrator, a position that introduced me to a bundle of challenges and obstacle involving humans as one of the crucial parameters. Plenty of critical decision making, budget and cost consideration, income generation, financial sustainability, management of change, organizational problem reporting and solving. All which have never been taught in any chemical engineering syllabus, as much as I am aware. I strongly believe this is something that needs to be seriously addressed and informed to all junior chemical engineers and engineering students.

All of us are born to be a leader. Earlier in life we are a leader to ourselves. Gradually later, we become a leader in our own family and small working surrounding. Moments later we will play a bigger leader role and soon we will discover that we become a leader in our society. This is parallel to our age, knowledge and experience. As we grow older, we have more knowledge and if we process it wisely, we will be a person full of wisdom.

My journey from 1999 to 2018 also witnesses another vital parameter that changes how everything works in this world, which is the transition from Industrial Revolution 3.0 to the advanced Industrial Revolution 4.0. The smart digital era creates an unimagined borderless world. Communication becomes swifter and decision makings are just in our palm. If God wills it, I will come out with the continuity of this book which will touch my overall process, journey from 1999 to 2018. I will keep you updated with the progress via my websites, Facebook Page and email list. Also the updates on my other publications
Stay Tuned.

Zaki Yamani Zakaria,
www.chem-eng.blogspot.com
www.facebook.com/ChemicalEngineeringWorld/
www.chem-eng.online
Skudai, Johor, October 2018.

BONUS: <u>FREE</u> Recommended Resources

Following are 5 very excellent resources (magazines) that <u>you can request at NO COST</u>. If you are interested with any of them, just click the magazine image or the request link and fill up your particulars. Personally, the magazines which will be digitally sent to us are very informative and provide high technical values that can build up our technical chemical engineering knowledge.

CHEMICAL ENGINEERING

Serves chemical engineering professionals in the chemical process industry including manufacturing, engineering, government, academia, financial institutions and others allied to the field serving the global chemical process industry.

Request for **Chemical Engineering** Magazine from:
http://chem-eng.online/chemical-engineering

HYDROCARBON PROCESSING

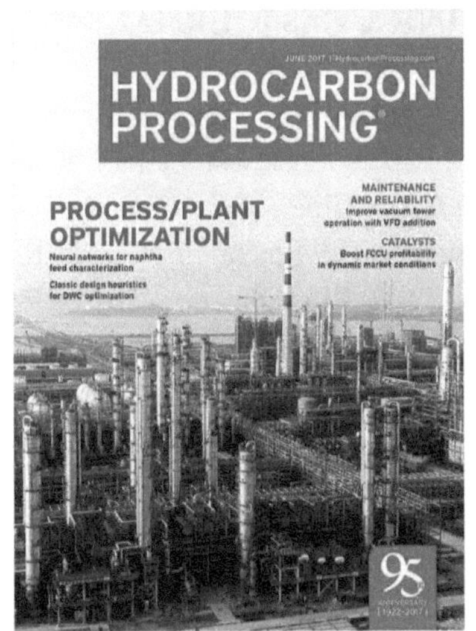

Since 1922, management and technical professionals throughout the world have turned to Hydrocarbon Processing for high-quality technical and operating information. Hydrocarbon Processing's editors and writers provide real-world case studies and practical information that readers can use to improve their companies' operations and their own professional skills.

Request for **Hydrocarbon Processing** Magazine from:

http://chem-eng.online/hydrocarbon-processing

PUMPS & SYSTEM

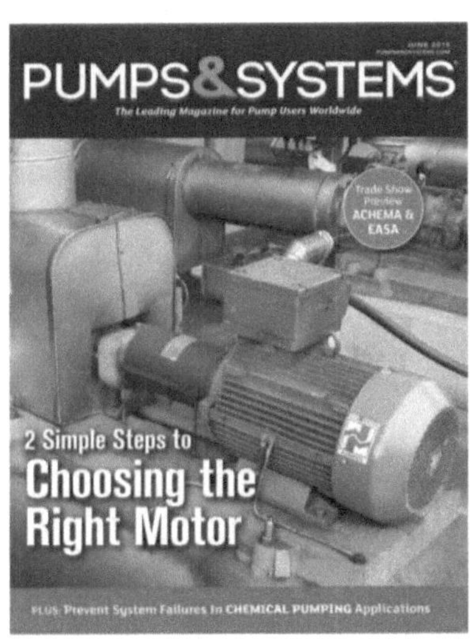

Pumps & Systems is the voice of the pump and rotating equipment industry. They deliver relevant industry news coverage and powerful technical information to managers, engineers, operators and maintenance professionals around the world.

Request for **Pumps & System** Magazine from:

http://chem-eng.online/pumps-system

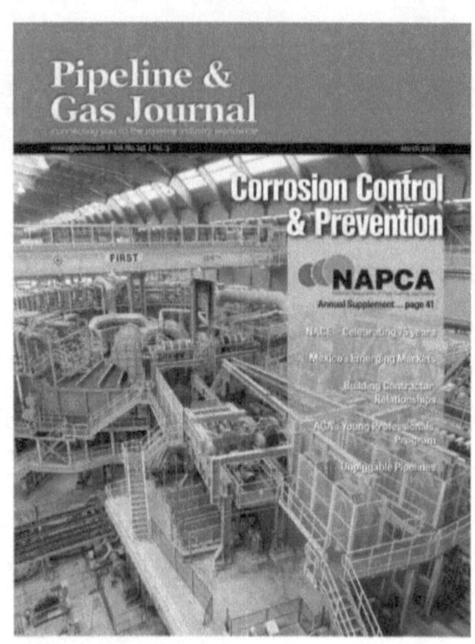

PIPELINE & GAS JOURNAL

Pipeline managers, engineers and operations personnel will benefit from this solid, reputable source of news and information on the technology, industry standards and "Best Practices" in oil & gas pipeline design, construction, operations, maintenance and integrity.

Request for **Pipeline & Gas Journal** Magazine from:
http://chem-eng.online/pipeline-gas-journal

INDUSTRIAL HEATING

Covers heat treatments, brazing, sintering, melting, process control, instrumentation, refractories, burners, heating elements, and other thermal processes typically in excess of 1000(degrees)F.

Request for **Industrial Heating** Magazine from:
http://chem-eng.online/industrial-heating